High Performance Concrete Optimal Composition Design

Leonid Dvorkin
Full Professor and Head
Department of Building Products Technology and Material Science
National University of Water and Environmental Engineering
Rivne,Ukraine

Vadim Zhitkovsky
Associate Professor
National University of Water and Environmental Engineering
Rivne, Ukraine

Oleh Bordiuzhenko
Associate Professor
National University of Water and Environmental Engineering
Rivne, Ukraine

Yuri Ribakov
Professor, Department of Civil Engineering
Ariel University, Ariel, Israel

CRC Press
Taylor & Francis Group
Boca Raton London New York

CRC Press is an imprint of the
Taylor & Francis Group, an **informa** business

Cover image source: Wirestock - Freepik.com

First edition published 2023
by CRC Press
6000 Broken Sound Parkway NW, Suite 300, Boca Raton, FL 33487-2742

and by CRC Press
4 Park Square, Milton Park, Abingdon, Oxon, OX14 4RN

Library of Congress Cataloging-in-Publication Data (applied for)

ISBN: 978-1-032-41386-0 (hbk)
ISBN: 978-1-032-41388-4 (pbk)
ISBN: 978-1-003-35786-5 (ebk)

DOI: 10.1201/9781003357865

Typeset in Times New Roman
by Innovative Processors

Preface

In modern construction, high performance concrete (HPC) is becoming increasingly important, allowing the production of highly responsible and unique structures. Development of the technology of this type of concrete requires in-depth and comprehensive studies of its properties and the possibility of their further improvement, expansion of the raw material base, including the expense of technogenic raw materials and industrial waste and development of a methodology for designing optimal compositions with a set of normalised quality indicators.

This monograph presents the results of the authors' research in the above direction in relation to the main variety of HPC – high-strength quick-hardening concrete – carried out at the National University of Water and Environmental Engineering (Ukraine) and Ariel University (Israel). The focus of the monograph is on the presentation of the developed methodology for multi-parameter prediction of the main properties of high-performance concrete and fibre-reinforced concrete based on it using experimental statistical models. On the basis of the obtained models, the possibility of designing optimal compositions of concrete and fibre-reinforced concrete mixes is considered.

This monograph complements the cycle of works performed earlier by the authors and covered in the following monographs (L. Dvorkin, V. Zhitkovsky and Y. Ribakov, *Concrete and Mortar Production Using Stone Sifting*, CRC Press, 2018, p. 158; Dvorkin L., Zhitkovsky V., Sonebi M., Marchuk V., Stepasiuk Y., *Improving Concrete and Mortar Using Modified Ash and Slag Cement*, CRC Press, 2020, p. 194; Dvorkin L., Zhitkovsky V., Lushnikova N., Ribakov Y., *Metakaolin and Fly Ash as Mineral Admixtures for Concrete*, CRC Press 2021, p. 240).

The authors are grateful to the team of scientists who took part in the research and to the engineers L. Matsko and H. Kyts, who assisted in the technical preparation of the book for publication.

The authors will be grateful for any feedback on the contents of the monograph and will try to take them into account in further research.

Leonid Dvorkin
Vadim Zhitkovsky
Oleh Bordiuzhenko
Yuri Ribakov

List of Symbols

W/C	–	Water-cement ratio
C/W	–	Cement-water ratio
f_{cm}	–	Compressive strength of concrete
P	–	Porosity
f_0	–	Non-porous material strength
d	–	Relative density
V_{air}	–	Air volume
W	–	Water consumption
C	–	Cement consumption
V_p	–	Pore volume
$V_{h.c}$	–	Volume of hydrated cement
α	–	Cement hydration degree
$d_{c.s}$	–	Density of cement stone
S	–	Fine aggregates consumption
C.A	–	Coarse aggregate consumption
W_s	–	Water demand of fine aggregates
$W_{c.a}$	–	Water demand of coarse aggregates
M_f	–	Fineness modulus
$K_{n.c}$	–	Cement paste normal consistency coefficient
NC	–	Cement paste normal consistency, %
HSC	–	High-strength concrete
HPC	–	High-performance concrete
R_c	–	Cement strength
$A (A_1)$	–	Quality of the raw materials coefficient
Sl	–	Standard cone slump, cm
Vb	–	Vebe index
$D_{c.a}$	–	Coarse aggregate largest size

$K_{c.e}$	–	Mineral additive of 'cementing efficiency' coefficient
$K_{w.s}, K_{w.c.a}$	–	Wetting coefficients of fine and coarse aggregates
V_a	–	Absolute volume of aggregate
δ	–	Thickness of cement paste layer
HSRHC	–	High-strength rapid-hardening cements
URHC	–	Ultra rapid-hardening cements
HSC	–	High-strength cements
FMMC	–	Fine-milled multicomponent cements
LWB	–	Low-water binders
SP	–	Superplasticizer
PFM	–	Polyfunctional modifiers
S_a	–	Specific surface area
MS	–	Microsilica
$f_{c.s}$	–	Cement stone compressive strength
WRE	–	Water-reducing effect
X_1-X_n	–	Designation of factors in experimental-statistical models
R_c	–	Cement strength
HMT	–	Heat-moisture treatment
CC	–	Composite cement
$f_{c,tn}$	–	Tensile strength at splitting
r	–	Proportion of sand in the mixture of aggregates
K_a	–	Grains moving apart coefficient
MPE	–	Mathematical planning of experiments
CS	–	Crushed stone consumption
ρ_b	–	Bulk density
ρ_{cs}	–	Crushed stone true density
ρ_s	–	Sand true density
$V_{c.p}$	–	Cement paste volume
V_a	–	Aggregate volume
V_{cs}	–	Crushed stone inter-granular voidage
k	–	Volume ratio of aggregate fractions coefficient
V_1-V_3	–	Factors indicating the volumetric ratio of components
P_1, P_2-P_m	–	Parameters of indicators of properties of concrete mixture and concrete
SCC	–	Self-compacting concrete
E_c	–	Modulus of elasticity
f_τ	–	Compressive strength of concrete at a certain duration of hardening (τ)
f'_{cm}	–	Cylindrical compressive strength of concrete
E_d	–	Dynamic modulus of elasticity

E_τ	–	Modulus of elasticity at the age of τ days
f_{pr}	–	Prism strength of concrete
$\varepsilon_{c.e}$	–	Conditional extensibility
m	–	Mass aggregate and cement ratio
$C_{m(28)}$	–	The limiting value of the measure of concrete creep when loaded in 28 days
M_{cr}	–	Modulus of cement stone cracking
f_{ctm}	–	Tensile strength
ε_{sh}	–	The magnitude of shrinkage deformations by the time cracks appear
F	–	Number of freeze cycles
P_k	–	Capillary porosity
SD	–	Degree of saturation
$V_{f.w}$	–	Volume of freezing water
W_f	–	Freezing water
K_t	–	Coefficient taking into account the freezing temperature
$P_{c.c}$	–	Conditionally closed porosity
K_F	–	Criterion of frost resistance
P_i	–	Integral porosity
F_c	–	Compensating factor
V_c	–	Volume of contraction pores in concrete
V_f	–	Volume of water in concrete that freezes at $-20°C$
K_c	–	Concrete compensatory factor
I	–	Impermeability
P2-P20	–	Concrete impermeability levels
$K_{f(\tau)}$	–	The concrete filtration coefficient over time when $\tau > 1$ day
V_r	–	Consumption of air-entraining admixture (Vinsol resin)

Contents

Part I
High-strength
Rapid-hardening Concrete

Theoretical Prerequisites

Influence of the Water-cement Ratio

Numerous studies and technological practices have proven that the main direction of increasing the strength of concrete is towards the reduction of water-cement ratio (W/C). This conclusion follows from the known physical dependence of the strength of solid bodies on their porosity. In general, this dependence can be expressed by an exponential function:

$$f_{cm} = f_0 (1 - P)^n \qquad (1.1)$$

where f_{cm} – compressive strength of concrete; P – porosity;
n – exponent that takes into account the peculiarities of the structure of materials;
f_0 – strength of non-porous material.

When porosity is replaced by the value of relative density $d = 1 - P$, Eq. (1.1) takes the form:

$$f_{cm} = f_0 \, d^n \qquad (1.2)$$

When calculating the strength of concrete depending on the relative density of the cement paste in the freshly prepared mixture, R. Feret took the indicator n equal to 2 [1]. O.E. Sheykin connects the strength of cement stone with its relative density, $n = 2.7$ [2]. According to S.M. Itskovich [3] for materials with a porous structure, n is approximately equal to 2, for granular materials, 3-6. There are data on the linear increase of n with increasing pore size [4].

The unambiguous relationship of W/C with the density of cement stone in concrete, characterised by the Fere parameter d, is based on the condition:

$$\frac{W + V_{air}}{C} = \frac{1}{d} - 1 \qquad (1.3)$$

where V_{air} is air volume; W and C are consumption of water and cement, respectively.

The density of cement stone $d_{c.s}$ at a constant W/C depends on the cement hydration degree (α) and its density (ρ_c) [5]:

$$d_{c.s} = \frac{1+0.23\alpha\rho_c}{1+\rho_c W/C} \qquad (1.4)$$

$$W/C = \frac{1+0.23\alpha\rho_c - d_{c.s}}{d_{c.s}\rho_c} \qquad (1.5)$$

Express the concrete pore volume by the known [5] dependence:

$$V_p = W - 0.23\alpha C \qquad (1.6)$$

where α is cement hydration degree, W and C are consumption of water and cement, respectively.

Dependence (1.6) assumes that dense aggregates are used, the pore volume of which can be neglected. Air volume in concrete of different origins is also not taken into account.

Volume of hydrated cement ($V_{h.c}$), according to Powers [6]:

$$V_{h.c} = 0.647\alpha C \qquad (1.7)$$

Then water consumption:

$$W = V_p^{c.s} + 0.23\alpha C \quad \text{or} \quad W = V_p^{c.s} + 0.35 V_{h.c} \qquad (1.8)$$

$$C = V_{h.c}/0.647\alpha \qquad (1.9)$$

$$W/C = 0.647\alpha\left(\frac{V_p^{c.s}}{V_{h.c}} + 0.35\right) \qquad (1.10)$$

where $V_p^{c.s}$ is pore volume of cement stone.

Thus, the water-cement ratio is directly proportional to the ratio of the pore volume of cement stone to the volume of hydrated cement at a constant value of α, the use of dense aggregates and the practical absence of entrained or residual air in the concrete mixture.

The total W/C of concrete can be represented [5] by the sum:

$$W/C = W_1/C + W_2/C \qquad (1.11)$$

where W_1/C is the water-cement ratio of the cement paste in concrete, due to the water directly bound by cement (W_1); W_2/C is water-cement ratio due to water (W_2) immobilised by aggregates:

$$W_2/C = \frac{W_S S}{C} + \frac{W_{C.A} C.A}{C} \qquad (1.12)$$

where W_S and $W_{C.A}$ are water demand of fine and coarse aggregates, respectively; S and $C.A$ are consumptions of fine and coarse aggregates.

The components in equation (1.11) are interdependent because on the one hand, W_1/C depends on the water demand of aggregates and on the other, the water demand indicator depends on the ratio of W_1/C to the normal consistency of cement paste $K_{n.c}$ [5].

In one of our works [7], we studied the comparative effect of W_1/C and W_2/C on the compressive strength of concrete at the age of 28 days (f_{cm}). For the production of concrete mixture, Portland cement with strength class 42.5 N and $K_{n.c} = 0.25$, quartz sand with a fineness modulus $M_f = 1.7$ and $M_f = 2.4$, granite crushed stone 5-20 mm were used ($W_{S_1}^0 = 0.081; W_{S_2}^0 = 0.065; W_{c.s}^0 = 0.022$), where W_S^0 and $W_{c.s}^0$ respectively, were the water demand of sand and crushed stone with normal cement consistency. Cement paste with different values of W_1/C was prepared. Then it was mixed with aggregates and water was added according to the condition (1.11).

At a constant W_1/C, an increase in W_2/C due to the transition to an aggregate with a higher water demand at constant cement consumption leads to an increase in the total W/C and a decrease in strength. The results of the experiments are given in Table 1.1.

To the greatest extent, the increase in W_2/C affects the strength of concrete at low values of W_1/C typical for high-strength concrete.

Thus, the W/C rule reflects the influence on the strength of concrete not only of the porosity of cement stone, but also to a large extent of the qualitative characteristics of its contact zone at the interface with aggregates.

The first calculation dependencies for the strength of high-strength concrete (HSC) were proposed by B.G. Skramtaev and Y.M. Bazhenov, who, based on the generally non-linear nature of the dependence of concrete strength on C/W, after extensive experimental research, proposed a formula for $C/W \geq 2.5$ [8, 9]:

$$f_{cm} = A_1 R_c (C/W + 0.5) \qquad (1.13)$$

where R_c – standard compressive strength of cement.

The values of the coefficient A_1 depend on the quality of the raw materials.

Materials for concrete	A_1
High quality	0.43
Ordinary	0.4
Reduced quality	0.37

By summarising the reference data, a single formula with averaged coefficient was derived:

$$f_{cm} = A R_c (C/W - 0.5) \qquad (1.14)$$

where $A = 0.56$.

Table 1.1: Influence of W_1/C and W_2/C on concrete strength (R_c = 52 MPa, C = 400 kg/m^3, $K_{n.c}$ = 0.25)

No.	W_1/C	$S,$ kg/m^3	$C.A,$ kg/m^3	W_s	$W_{c.a}$	W_2/C	W/C	f_{cm} MPa
		$W_{S_2}^0$ =0.081; $W_{c.a}^0$ =0.022						
1	$0.87K_{n.c}$	650	1150	0.027	0.007	0.064	0.283	94.6
2	$K_{n.c}$	700	1100	0.081	0.022	0.202	0.452	53.4
3	$1.1K_{n.c}$	750	1100	0.095	0.028	0.255	0.530	43.3
4	$1.2K_{n.c}$	800	1100	0.110	0.031	0.305	0.605	35.9
5	$1.3K_{n.c}$	800	1100	0.121	0.033	0.333	0.658	31.8
		$W_{S_2}^0$ =0.065; $W_{c.a}^0$ =0.016						
6	$0.88K_{n.c}$	650	1150	0.019	0.006	0.048	0.267	101.2
7	$K_{n.c}$	700	1100	0.065	0.016	0.158	0.408	60.9
8	$1.1K_{n.c}$	750	1100	0.077	0.024	0.21	0.485	48.7
9	$1.2K_{n.c}$	800	1100	0.081	0.029	0.242	0.552	41.0
10	$1.3K_{n.c}$	800	1100	0.085	0.031	0.255	0.580	38.2

Note: W_s^0 i $W_{c.a}^0$ are indicators of water demand of aggregates found according to the standardised method [10].

V.P. Sizov suggested using formula 1.14 in the entire range of concrete strength values for the design of concrete compositions, while taking the value of coefficient A according to Table 1.2 and additionally specified by corrections ΔA_i [11].

Amendments are taken into account:

- ΔA_i – standard cone slump (Sl) –

$$\Delta A_1 = -0.0033\ Sl \qquad \text{(at } Sl \leq 20 \text{ mm } \Delta A_1 = 0) \qquad (1.15)$$

- Vebe index (V_b) –

$$\Delta A_1 = 0.0012\ Vb \qquad \text{(at } Vb \leq 5 \text{ sek. } \Delta A_1 = 0) \qquad (1.16)$$

- ΔA_2 normal consistency of cement (NC)

$$\text{at } NC > 27\% \qquad \qquad \Delta A_2 = -0.0075\ (NC\text{-}27) \qquad (1.17)$$

$$\text{at } NC < 27\% \qquad \qquad \Delta A_2 = 0.01\ (27\text{-}NC) \qquad (1.18)$$

- ΔA_3 sand fineness modulus (M_f)

$$\Delta A_3 = 0.01\ (M_f\text{-}3) \qquad (1.19)$$

- ΔA_4 the largest size of coarse aggregate $D_{c.a}$ – at $D_{c.a}$ = 10 mm ΔA_4 =–0.03; $D_{c.a}$ = 20 mm ΔA_4 = –0.02; $D_{c.a}$ = 40 mm ΔA_4 = –0.01.

When using conditioned aggregates recommended use of the formula [12] in the entire C/W interval:

$$f_{cm} = \frac{(2.3R_c + 100)C/W - 80}{10} \tag{1.20}$$

This formula has limited application and does not take into account the quality of aggregates.

Table 1.2: The value of coefficient A (following [5])

Aggregate type	The content of clay, dust and silt in crushed stone (gravel) and sand %	The value of coefficient A for concrete at		
		Crushed stones	Gravel mountain	River and sea gravel
Crushed stone (gravel) sand	0 0	0.64	0.6	0.57
Crushed stone (gravel) sand	0 3	0.61	0.56	0.53
Crushed stone (gravel) sand	1 3	0.58	0.53	0.5
Crushed stone (gravel) sand	2 3	0.55	0.5	0.47
Crushed stone (gravel) sand	2 5	0.52	0.47	0.44

Taking into account the results of research on high-strength concrete for calculating their strength, Y.M. Bazhenov proposed a formula [9]:

$$f_{cm} = KARc\,(C/W - 0.5) \tag{1.21}$$

where K – coefficient that takes into account the specifics of the effect of a chemical admixture on the strength of concrete at a certain age.

The given formulas are valid for concrete made of moderately stiff and flow concrete mixtures, placed by vibration, with a compaction coefficient of not less than 0.98.

To more fully take into account the factors that affect the strength of concrete, the coefficient A can be expressed as the product $pA = A \cdot A_1 A_2 - A_n$ where $A_1 - A_n$ are additional coefficients that take into account the effect of temperature, duration of hardening, admixtures, etc. on the strength of concrete.

The predictive ability of calculation formulas for the strength of concrete increases significantly when using the 'modified' C/W [5]:

$$(C/W)_m = \frac{V_c + K_{c.e}V_{ad}}{W + V_{air}} \tag{1.22}$$

where $K_{c.e}$ – coefficient of 'cementing efficiency' or 'cement equivalent' of the mineral admixture introduced into the concrete mixture to save cement; V_c, V_{ad}, V_{air} are absolute volumes of cement, mineral admixtures and air.

Use of the $(C/W)_m$ parameter opens up the possibility of developing fairly simple universal methods for calculating the compositions of normal-weight and light concrete, based on the same physical prerequisites.

The concrete strength formula depending on $(C/W)_m$ has the general form:

$$f_{cm} = pAR_c ((C/W)_m + b) \tag{1.23}$$

where pA_i is multiplicative coefficient ($pA_i = A_1, A_2\text{-}A_n$), which takes into account a number of technological factors that affect the strength of concrete at a constant $(C/W)_m$, age of concrete, temperature and humidity conditions of hardening, etc.

The use of the 'modified' C/W is rational, especially for the calculation of concrete compositions with limited cement consumption when introducing dispersed mineral admixtures. Taking into account the pore volume of aggregates, it is also effective in calculating the composition of lightweight concrete and the volume of entrained air – concrete with air-entraining admixtures. In the absence of mineral admixtures in concrete, the formula (1.23) is transformed into the formula of the usual form.

Water Demand and Workability

Water demand and workability are the most important interconnected technological properties of concrete mixtures, which determine both their ability to compact and, to a large extent, the properties of hardened concrete.

Not taking into account the amount of hydrated water, which is insignificant before the concrete hardens, can write the equation of the water balance of the concrete mixture:

$$W = XK_{n.c}C + K_{w.s}S + K_{w.c.a}C.A + W_{ad} + W_h \tag{1.24}$$

where W is water content, due to the required workability of concrete mix, kg/m³; C, S and $C.A$ are consumptions of cement, sand and crushed stone (gravel), respectively, kg/m³; $K_{n.c}$, $K_{w.s}$, $K_{w.c.a}$ are normal consistency of cement and wetting coefficients of fine and coarse aggregates; $X = (W/C)_p/K_{n.c}$ is the relative moisture content of the cement paste in the concrete mixture, where $(W/C)_p$ is the water-cement ratio of cement paste; W_{ad} is water adsorbed by pores of aggregates, kg/m³; W_h is water mechanically held in the pore space between aggregate grains covered with cement paste, kg/m³.

The minimum amount of water required for the mixtures being formed is approximately equal to:

$$W_{min} = X_{m.c} K_{n.c} C + K_{w.s}S + K_{w.c.a} C.S + W_{ad} \tag{1.25}$$

where $X_{m.c}$ is relative moisture content of cement paste, which corresponds to the maximum moisture content of cement, at which it practically does not contain capillary moisture in the cells between the watered grains ($X_{m.w} \le 0.876$).

Wetting coefficients $K_{w.s}$ and $K_{w.c.a}$ characterise the specific amount of water retained by fine and coarse aggregates, respectively, in a film state on their surface. They depend on the amount of surface energy, coarseness and relief of the surface of aggregate grains. Various methods of determining wetting coefficients are proposed – testing directly moistened sand and crushed stone (gravel) or cement mortar (concrete) with a certain consistency of cement paste. At a certain moisture content of the sand, the electrical resistance changes sharply, indicating a decrease in the binding energy between water and sand after its complete wetting.

The wetting coefficient of quartz sand, taking into account water absorption, depending on the size of the fraction, ranges from 0.72 (5-2.5 mm) to 5.04% (0.3-0.15 mm) and granite crushed stone from 1.21 (5-10 mm) to 0.75% (40-60 mm) [13]. Wetting ends with equilibrium, in which the concrete mixture acquires a loose-earthy state. At the same time, the estimated *W/C* of such a mixture (when $W_{ad} = 0$) is equal to:

$$(W/C)_c^0 = 0.876 K_{n.c} + \frac{K_{w.s}S + K_{w.c.a}C.A}{C} \tag{1.26}$$

It is not difficult to calculate that the value $(W/C)_c^0 = 0.25$-0.35.

With intense mechanical effects, for example, pressing or vibro-pressing, part of the wetting water is squeezed out and $(W/C)_c^0$ is reduced. The value of $X_{m.w}$ at a pressure of 50 MPa decreases to almost 0.1. In hot-pressed samples of cement stone, Roy and Gouda reached *W/C* = 0.093 [5]. The total volume of wetting water on the surface of the aggregate, as the pressing pressure increases, can approach the adsorption volume. A significant reduction in $X_{m.c}$ can also be achieved due to the use of effective superplasticizers. This is proven by the practice of using low water-consumption binders in concrete and mortar, obtained by joint grinding of clinker and mineral admixtures with the introduction of superplasticizers.

At about the same time (in the early 1930s) and independently of each other, V.I. Soroker in the USSR and F.R. McMillan in the USA established the rule of constancy of water demand (CWD). They established that with constant water content, cement consumption in the range of 200-400 kg/m³ does not significantly affect the workability of concrete mixtures.

A significant stabilisation of the upper limit of the CWD area and taking into account the features of the cements used is achieved when expressing it through the critical *W/C* ($(W/C)_{cr}$), which is equal to $1.68 K_{n.c}$ on average, where $K_{n.c}$ is the *W/C*, which corresponds to the normal consistency of cement paste.

From a physical point of view, the rule of constancy of water demand is that with an increase in *C/W* to a certain critical value, the increase in the structural viscosity of the cement paste in the concrete mix is compensated by an increase in its quantity and, accordingly, the thickness of the cement paste layer on the aggregate grains. Beyond the critical *C/W*, increasing the amount of cement 'lubricant' no longer compensates for the progressively increasing water demand of the concrete mixture.

B.G. Skramtaev and Y.M. Bazhenov proposed integral quantitative indicators of water demand of aggregates determined by comparative tests of cement paste, mortar and concrete mixture. These indicators determine the amount of water that must be added to the cement paste per unit mass of sand or crushed stone (gravel), respectively, in order to obtain a mortar mixture of the composition 1:2 or a concrete mixture of the composition 1:2:3.5 with the same fluidity after 30 minutes, after kneading as a paste of normal consistency.

Taking into account the indicators of water demand of fine (W_s) and coarse aggregates ($W_{c.a}$) is convenient for their comparative assessment since, unlike the modulus of fineness and specific surface area, it allows for a generalised assessment of the features of sand or crushed stone (gravel) that affect the water content of concrete mixtures.

In the general form, W_s and $W_{c.a}$ can be found using the equations:

$$W_s = K_{w.s} W_s^0, \quad W_{c.a} = K_{w.c.a} W_{c.a}^0 \tag{1.27}$$

where $K_{w.s}$, $K_{w.c.a}$ are coefficients, the values of which, as our experiments have shown, depend on the parameter

$$X = (W/C)_{c.p}/K_{n.c}, \ W/C_{c.p} \text{ is cement paste } W/C.$$

At $X = 0.876$, the W_s indicator is almost equal to the wetting coefficient, $X = 1$ is the water demand indicator, which is determined according to the method of B.G. Skramtaev and Y.M. Bazhenov. At $X = 1.65$, the cement paste reaches its water-retention capacity and the water demand of the aggregate becomes almost maximal.

Concrete mixtures must maintain homogeneity during transportation, laying and compaction, and not delaminate.

An attempt to calculate the total amount of water retained by the concrete mixture (W_r) without significant water separation was made by I.M. Hrushko [14]. He proposed the equation:

$$W_r = 1.35 K_{n.c} C + W_s S + 0.07 S_{c.a} \tag{1.28}$$

where W_s is water demand of sand; $S_{c.a}$ is specific surface area of coarse aggregate, m^2/kg.

The maximum possible amount of water retained by the concrete mixture can also be represented in the form of an equation:

$$W_r = 1.35\text{-}1.65 \ K_{n.c} C + W_s S + W_{c.a} C.A \tag{1.29}$$

where W_s and $W_{c.a}$ are water demand of fine and coarse concrete aggregates when $X = 1.35\text{-}1.65$ and using cement with the specified value $K_{n.c}$.

In view of the crucial influence of water content on the indicators of workability of concrete mixes, a number of relevant equations are proposed. So, V.V. Mykhaylov [15] proposed a linear equation connecting the water content of the concrete mixture (W) and the workability it estimated by the stiffness index (St):

$$W = 0.5(60 - St) + W_0 \qquad (1.30)$$

where W_0 is water content of the mixture at 60 seconds of its workability (according to a technical viscometer).

A linear dependence is also proposed [16] for the relationship between slump of cone (Sl, cm) and the content of 'free' or 'mobile' water in the concrete mixture (W_f):

$$Sl = nW_f \qquad (1.31)$$

where n – empirical coefficient.

Linear dependencies, however, are valid only within narrow limits of the change workability of concrete mix. They do not take into account the rule of constancy of water demand in a certain range of C/W.

The analysis of numerous experimental data shows that the water consumption of the concrete mixture (W_0) is related to the indicator of workability (cone slump) and the size of the coarse aggregate by a quadratic polynomial function.

With a change in the slump of the cone in the range of $Sl = 0\text{-}20$ cm and the maximum coarseness of the aggregate from 5 to 80 mm, the water consumption can be approximately calculated according to the equation:

$$W_0 = 176 - 0.8D_{c.s} + 6.1Sl + 0.0029D_{c.s}^2 - 0.14Sl^2 \qquad (1.32)$$

where $D_{c.s}$ is the largest size of crushed stone, mm.

To take into account the characteristics of the starting materials, the consumption of cement and the temperature of the concrete mixture, the calculated values of W_0 can be adjusted according to known recommendations (Table 1.3).

Other corrections can be introduced into the calculation of water consumption if appropriate experimental material is available, but the desire to take into account the influence of all factors on water consumption, as well as the composition, is unproductive. The final correction of the really necessary water consumption to achieve the given indicator of workability should be performed in the process of production adaptation of concrete mixtures compositions.

The task of calculating water consumption includes taking into account the effect of plasticizing admixtures, if such are introduced.

To determine the change in water consumption DW of concrete mixtures beyond the rule of constancy of water consumption, which applies to the critical W/C $(W/C)_{cr}$ equal is an average 1.68 $K_{n.c}$ [5], can use the empirical formula:

$$\Delta W = \left(C/W - \frac{1}{1.68K_{n.c}} \right) \left(\frac{W_0}{100} \right)^{5.5} \qquad (1.33)$$

where W_0 is water demand of concrete mixture at $W/C < (W/C)_{cr}$.

Table 1.3: Amendments for adjusting the water consumption of the concrete mixture

No.	Factors taken into account	Amendments ΔW, l/m^3
	I: Type of Coarse Aggregate	
1	Crushed stone from metamorphic and sedimentary rocks with compressive strength 40-80 MPa	$\Delta W_1 = W_0 + (4\text{-}13)$
2	Mountain gravel	$\Delta W_2 = W_0 - (5\text{-}10)$
3	Sea and river gravel	$\Delta W_3 = W_0 - (9\text{-}15)$
4	Crushed stone from rocks with a smooth fracture surface (diabase, basalt, etc.)	$\Delta W_4 = W_0 - 3$
5	Washed crushed stone	$\Delta W_5 = W_0 - 6$
6	Content of silt and dust in crushed stone in % over 1% (X_1) and particles smaller than 5 mm (X_2) over 5%	$\Delta W_6 = W_0 + X_1$ $\Delta W_6' = W_0 + 2X_2$
	II: Coarseness and Contamination of Sand	
7	Change of sand module fineness for every 0.5 less than 3(X_3) more than 3(X_4)	$\Delta W_7 = W_0 + (3\text{-}5)X_3$ $\Delta W_7' = W_0 - (3\text{-}5)X_4$
8	Sand with a smooth, well-rounded surface	$\Delta W_8 = W_0 - 4$
9	Washed sand	$\Delta W_9 = W_0 - 7$
10	An increase in the content of silt and dust in the sand for every % more than 3% (X_5)	$\Delta W_{10} = W_0 + 2X_5$
11	Change in the normal consistency of cement paste in % more than 28% (X_6), less than 28%	$\Delta W_{11} = W_0 + 4X_6$ $\Delta W_{11}' = W_0 - 4X_6$
12	Change in cement consumption for every 10 kg over 350 kg/m³ (X_7)	$\Delta W_{12} = W_0 + X_7$
	III: The Temperature of the Concrete Mixture, °C	
13	5	$\Delta W_{13} = W_0 - 5$
14	10	$\Delta W_{13} = W_0 - 4$
15	15	$\Delta W_{13} = W_0 - 2$
16	25	$\Delta W_{13} = W_0 + 3$
17	30	$\Delta W_{13} = W_0 + 7$
18	35	$\Delta W_{13} = W_0 + 11$

Note: W_0 is estimated water consumption without corrections.

Effect of the Ratio of Concrete Aggregates

One of the important quality indicators of aggregates is their grain (granulometric) composition. The first works on the design of the grain composition of concrete aggregates were aimed at ensuring minimal voids in mixtures of grains of different shapes and sizes.

With known values of the bulk density of the aggregate (ρ_b) and the density of its grains (ρ_g), the calculated value of the emptiness (E^0):

$$E^0 = \left(1 - \frac{\rho_b}{\rho_g}\right) \tag{1.34}$$

When mixing aggregates, the calculated emptiness value can be determined by the formula:

$$E^0 = 1 - \frac{\rho_{b.a}}{m_a}[V_1(1 - E_1^0) + V_2(1 - E_2^0) + \ldots + V_n(1 - E_n^0)] \tag{1.35}$$

where m_a and $\rho_{b.a}$ are the mass of the aggregate mixture and its bulk density; V_1, V_2-V_n are bulk volumes of fractions that are mixed; E_1^0, $E_2^0 \ldots E_n^0$ – emptiness fractions of mixed aggregates.

In practice, the most and least dense arrangement of aggregates grains is unlikely. Void values increase with increasing flakiness of grains, especially when using elongated grains.

Two approaches have been developed to ensure dense mixtures of aggregate grains: the choice of discontinuous or continuous in their composition.

Proponents of discontinuous granulometry of the aggregate mixture proceed from the geometric regularities of the stacking of granular materials. Although the discontinuous grain composition provides less voids in the grain mixture, most researchers prefer the continuous grain composition of aggregates in concrete mixtures. This is explained by the need for the same workability of concrete mixtures in the latter case of a smaller volume of small fractions and, accordingly, the consumption of cement for coating the grains. In addition, mixtures with a continuous grain composition are less prone to delamination.

Various 'ideal' sieving curves have been proposed to select a continuous grain composition of aggregates, of which Fuller's, Bolomey's, and Hummel's curves, which are expressed by quite close formulas, have gained the greatest popularity.

For real aggregates, the grain composition always deviates from the ideal curve; therefore, in the standards defining the requirements for coarse and fine aggregates, the recommended range of grain compositions is indicated, going beyond which leads to a significant overconsumption of cement.

For plastic concrete mixtures, the thickness and consistency of the layer of cement paste have a significant influence on the optimal ratio of aggregate fractions, which change, depending on the specified values of the properties of the concrete mixture and concrete. From the point of view of minimising the consumption of cement, it is important that the composition of the aggregate ensures the minimum possible voids with the smallest total surface.

The required volume of cement paste for the production of dense concrete:

$$V_{c.p} = E_a^0 \, S_a + \delta S_a V_a + V_{air} \tag{1.36}$$

where E_a^0 and S_a are emptiness (by absolute volume) and specific surface area of the aggregate; V_a is absolute volume of aggregate; V_{air} is air volume; δ is the thickness of the layer of cement paste.

The task of choosing the optimal ratio of aggregates in the concrete mixture has subsequently undergone a certain evolution. Many researchers have proven that by changing the volume ratio of fine and coarse aggregates with a constant volume of cement paste, it is possible to achieve the best workability or with constant workability of the concrete mixture of the highest concrete strength. At the same time, the optimal value of the volume ratio of aggregates, as a rule, does not coincide with the ratio that ensures the minimum voidness of their mixture.

At present, there are sufficiently tested recommendations for determining the proportion of sand in a mixture of fine and coarse aggregate, established empirically, usually with the condition of achieving the best workability of the concrete mixture at a given volume of cement paste and W/C.

V.N. Shmygalsky established the 'consistency constancy rule' [17], according to which, given the W/C and quality of the components, the workability (consistency) of the mixture with a change in the proportion of fine aggregate in the total volume of aggregates (value r) practically does not change, if the conventional thickness of the cement paste shell on aggregate grains δ remains constant.

As a consequence of the rule of constancy of the consistency of concrete mixtures, the rule of optimal sand content can be considered. For most of the compositions of concrete mixtures, the optimal content of sand in the mixture of aggregates (r_o) is found from the condition: $r \rightarrow r_o$ at constant volume of cement paste $V_{c.p}$ = constant if the viscosity of the concrete mixture is $\eta_{c.m} \rightarrow$ min. At the same time, the water content is accordingly minimised.

Influence of the Degree of Cement Hydration

Already the first dependencies for the strength of cement stone and concrete took into account the influence of the degree of cement hydration. In particular, Powers proposed the parameter X, which is uniquely related to the strength of cement stone:

$$X = \frac{K_g V_{s.c} \alpha}{V_{s.c} + W/C} \approx \frac{0.47\alpha}{0.319\alpha + W/C'} \tag{1.37}$$

where K_g = 2.09-2.2 is coefficient of increase in the volume of hydration products (gel); $V_{s.c}$ is specific volume of cement $V_{s.c}$ = $1/\rho_c$ = 0.319 cm^3/g – the value is the inverse of the cement density (ρ_c); α is part of the cement that has undergone hydration (degree of hydration).

The degree of hydration of cement of ordinary chemical and mineralogical composition is also proposed to be calculated using the empirical equation:

$$\alpha = k \lg \tau - B \qquad (1.38)$$

where k is reaction rate constant; τ is duration of hardening, days; B is constant that reflects the duration of the induction period of hydration.

The current state of concrete science offers a certain system of calculation dependencies that allow taking into account the influence of the main structure-forming factors on the main parameters of the composition of concrete mixtures, water consumption and water-cement ratio, but when using multicomponent concrete mixtures and the need to take into account the specific features of the initial components and technological parameters, these calculation dependencies need to be adjusted.

Technological Ways for Producing High-strength Rapid-hardening Concrete

The range of concrete types used in modern construction is rather wide and constantly expanding. This is mainly due to scientific and technological progress in technology of binders, aggregates and various admixtures for concrete. In recent decades, a new generation of high-strength concrete – High-performance concrete (HPC) is increasingly used in both structural and special concrete [18]. These concretes are obtained, based on concrete mixtures at high workability, providing high strength both at design and at an early age, and characterised by volume stability, low abrasion, high impermeability, chemical resistance, frost resistance, bactericidal, fungicidal and other characteristics, which meet national and international standards. Currently, technology of various types of HPCs is being actively introduced – High-strength concrete (HSC), Self-compacted Concrete (SCC), Reactive Powder Concrete (RPC), High-performance fibre-reinforced concrete, etc. Producing such concretes that have a complex of unique properties became possible due to advanced technological decisions and, first of all, using modern organic and mineral admixtures including superplasticizers, highly active mineral and other admixtures.

To date, there is no single definition for concrete that can be classified as high-strength one. The conditional border between ordinary and high-strength concrete changes with the development of concrete technology. In the 50s of the last century, high-strength concrete was classified as C25-C40, in the 60s above C50-C60. Currently, high-strength concrete has at 28 days compressive strength of 70-150 MPa. The European standard EN206 [19] provides a possibility of producing and application of concrete including class C115. Thanks to application of effective superplasticizers and microsilica admixtures, industrial technology of producing concrete with strength in specified range was developed and corresponding normative documents were prepared [20-23]. Such concrete is increasingly used for structural elements of monolithic frames in high-rise buildings, bridges, offshore platforms, vibro-hydropressed pipes, etc. In laboratory conditions concrete with a strength of up to 200 MPa and higher was obtained [23].

Achieving high strength of normal weight concrete is possible by increasing the density and strength of cement stone (cohesion factor) and the strength of the contact zone (adhesion factor) [5]. Development of concrete science in the late 20th and early 21st centuries allowed determination of the main ways to obtain high-strength rapid-hardening concrete. These ways can be arranged in the following order:

1. Application of high-strength cements and high-quality aggregates.
2. Reduction of W/C first of all at the expense of using effective plasticizing admixtures.
3. Regulating the hydration processes and concrete structure formation by a complex of effective technological solutions and first of all, highly active mineral admixtures.

High-strength Cements

Four main directions of obtaining high-strength cements have been developed [24]:

- Consistent optimisation of all technological operations of cement production;
- Changing the composition of clinker minerals and their alloying by introducing special admixtures into the raw material mixture;
- Addition into cement special crystallisation additive;
- Synthesis of mixed cements, each of which components has a positive effect on the hydration hardening of the other components.

Fine grinding and high homogeneity of the raw materials mixture, strong and uniform firing of clinker, correct selection of the type and ash content of fuel, sharp cooling of clinker are of great importance for obtaining high-strength rapid-hardening cements (HSRHC). A great influence on the clinker activity is also its microcrystalline structure. Cement of improved quality is obtained mainly at correct crystallisation of alite, which is characteristic of the so-called monadoblastic structure, which is formed under optimal burning conditions. Early strength of Portland cement is determined by the grain content of less than 10 μm, and strength at later hardening ages up to 30 μm. For HSRHC, the content of cement fractions smaller than 30 μm should be at least 65-75% and depending on the Portland cement class, it can reach 80%.

High-strength ultra-rapid hardening cements (URHC) were obtained by significant increase in the clinker component of the tricalcium silicate content and optimisation of the cement grain composition. The recommended grain composition of URHC [24], is given in Table 2.1.

Table 2.1: Grain composition of URHC

Cement compressive strength (28 days), MPa	Specific surface, m^2/kg	The content of particles, % (particle size in microns)			
		<5	5-30	30	>60
60	About 300	5-12	30-50	30-45	5
70	370-400	20-28	45-70	11-25	6
80	About 600	15-30	55-75	3-25	1

Research results have shown that when the specific surface area of cement increases from 200 to 600 m^2/kg at the optimal gypsum content for each dispersion level, the hydration degree and strength at one to three days increases significantly, the strength at 28 days increases only to a certain limit and then decreases. Providing the required grain composition of HSRHC is possible during grinding in mills operated in a closed cycle. At the same time, however, there is a significant (up to 40-50%) decrease in the grinding unit's productivity.

Some countries have developed URHC which allows achievement of compressive strength of 5-20 MPa at two to six hours after mixing, 70-80 MPa and more at 28 days [24]. Such cements include fluorine-containing ultra-rapid hardening Portland cement. Raw materials for this cement use limestone, slags of secondary re-melting of aluminum, calcium fluoride and a special admixture for maintenance of fluorine in clinker. Fluorine-modified cement achieves compressive strength of 5-8 MPa after six hours of normal hardening. The use of fluorine-containing cement enables provision of the necessary stripping strength in the production of reinforced concrete products after one hour of heat treatment.

A promising method of obtaining URHC is by adding calcium sulfates in the amount of about 10% into the raw material mixture. Calcium sulfoaluminates formed during clinker burning provide the cement rapid-hardening ability and high-hydraulic activity. The raw material component for production of clinker for such cement can be phosphogypsum and other industrial wastes [24].

Promising types of binders for high-strength concrete include low-water binders (LWB) and fine-milled multicomponent cements (FMMC). Development of theoretical ideas on cements with mineral admixtures and surface-active admixtures, as well as using effective superplasticizers for cement systems enabled to obtain a new generation of composite binders, characterised by low water demand and high strength at a relatively low mass ratio of clinker and mineral admixtures [24].

LWB are prepared by fine grinding of clinker or ready-made Portland cement and active mineral admixtures in the presence of a powdered superplasticizer. Distinctive features of LWB are high dispersion ($S = 400$-500 m^2/kg), low water demand (normal consistency up to 18%), high strength (100 MPa and more) [25]. When the binder is finely ground, mechano-chemical activation takes place and as a result, the number of active centres and new free valencies per unit volume of clinker and mineral filler grains increase. As a result, the plasticizing effect of admixtures and the binder strength significantly increase [24].

FMMC are hydraulic binders obtained by common fine grinding of Portland cement clinker or Portland cement and active or inert mineral admixtures. The total weight content of admixtures in cements of this type can be 20-50% or more. The optimal dispersion of FMMC averages 450 m^2/kg [24].

Production of FMMC can be organised both at cement plants and at the construction industry enterprises with the use of ball, jet mills or vibromills. FMMC is used in concrete mixtures and mortars when adding with water-mixing superplasticizing admixtures [26]. In this case, considering the significant number of mineral admixtures, the content of superplasticizers, respectively, increases in comparison with that required for admixture-free Portland cements.

Along with special high-strength cements, technology of which has been tested at cement plants, but not sufficiently implemented considering the technological difficulties in obtaining them, locally available conventional cements can be used for producing high-strength concrete.

Aggregates for High-strength Concrete

For high-strength concrete choosing aggregates, in addition to strength, their density, porosity and water absorption are taken into account. Crushed stone or gravel, with an average grains density of at least 2.5 g/cm^3 and a water absorption of up to 0.5% for igneous and metamorphic rocks and 1% for sedimentary rocks, is used. For less responsible concrete, the grain density of coarse aggregate should be at least 2.3 g/cm^3 and water absorption up to 0.8% for aggregate from igneous and metamorphic rocks and 2% for sedimentary rocks [24]. For normal-weight concrete aggregates, the density of the rocks, which characterises their weight per unit volume in an absolutely dense state, varies within narrow limits (Table 2.2) and is taken into account when calculating their porosity and concrete mixture's composition.

The average density of rocks depends on their porosity and varies widely. For rocks of a certain mineral composition and structure, this indicator can serve as an indirect characteristic of their strength and durability. Porosity of igneous rocks usually does not exceed 1.5%; for sedimentary, especially carbonate rocks it reaches 40%. For production of crushed stone used in normal-weight concrete, carbonate rocks with porosity in the range of 5-15 are used. The open porosity of rocks is characterised by their water absorption. Water absorption of igneous rocks that have not undergone weathering processes usually does not exceed 0.7% and for sedimentary rocks, it reaches 10% and more.

For producing concrete mixture, it is advisable to use coarse aggregate in the form of separately dosed fractions: 5-10 (3-10); 10-20; 20-40; 40-80; 80-120 mm. The use of coarse aggregates in the form of a mixture of adjacent fractions is also allowed. The fraction of 3-10 mm is used in case of application of fine aggregate sands with a fineness modulus up to 2.5. The content of individual fractions in

Table 2.2: Basic properties of rocks used for HSC

Rocks	Density, g/cm³	Compressive strength, MPa	Modulus of elasticity, 10⁴ MPa	Frost resistance cycles
Igneous intrusive:				
Granites	2.53-2.7	100-260	5-8	100-300
Gabbro	2.85-3.05	100-350	9-11	100-300
Effusive:				
Porphyries	2.54-2.66	60-150	6-8	50-200
Basalt	2.22-3.07	110-500	8-8.3	50-200
Metamorphic:				
Gneiss	2.0-2.5	10-200	6-7	25-200
Quartzite	2.55-2.7	100-250	7-9	100-300
Sedimentary:				
Carbonate	1.7-2.7	5-200	0.2-9	up to 300
Sandstones	2.0-2.5	10-250	1.4-5	15-300

coarse aggregate in concrete composition is set according to the condition of achieving its highest density and is within the limits indicated in Table 2.3. The quality of aggregates is significantly affected by the content of dusty, clayey and silty impurities. Particles larger than 0.005-0.05 mm are classified as dusty, while clay and silty particles are up to 0.005 mm in size. The limitation of the de-silted impurities content in aggregates is due to the negative effect of the films formed by them on the cement-stone adhesion with aggregates, and, as a result, on the strength, frost resistance and other properties of concrete as well as on concrete mixtures water demand.

Table 2.3: Recommended composition of the coarse aggregate fractions mixture

Maximum aggregate size, mm	The content of fractions in the coarse aggregate, %				
	5 (3)-10 mm	10-20 mm	20-40 mm	40-80 mm	80-120 mm
10	100	-	-	-	-
20	25-40	40-75	-	-	-
40	15-25	20-35	40-65	-	-
80	10-20	15-25	20-35	35-55	-
120	5-10	10-20	15-25	20-30	30-40

When establishing the permissible amount of dust-like and clay impurities in coarse aggregate, the type of rock is taken into account: for igneous and metamorphic rocks in concrete of all classes up to 1%, for sedimentary rocks for concrete classes C20/25 and higher – 2%, for C15/20 and below – 3% by weight.

The necessary frost resistance of crushed stone and gravel is normalised, taking into account the average monthly temperature of the coldest month in the year. If the latter ranges between 0-10°C, the frost resistance class of crushed stone and gravel should be at least F100, from –10 to –20°C – F200, below –20°C – F300.

For sand, which is used as a fine aggregate of concrete, the determining quality features are the grain composition, content of dusty and clay (de-silted) particles and petrographic characteristics. If the grain composition of natural sand does not meet the recommended requirements, corrective admixtures are used. For fine and very fine sand, it can be enlarging admixtures – coarser sand, including sand from natural stone crushing sifting. For coarse sands, if it is necessary to adjust their grain composition, sands with a lower fineness modulus are used.

Harmful impurities in concrete aggregates that cause its corrosion and surface quality deterioration include: amorphous varieties of silicon dioxide (chalcedony, opal, flint, etc.); sulphur, sulphides, sulphates; magnetite, iron hydroxides. Concrete strength and durability are reduced by such impurities in aggregates as coal, graphite, combustible shale and apatite, nepheline, phosphorite. Some impurities that contain water-soluble chlorides, sulphur, sulphides and sulphates can cause corrosion of concrete reinforcement.

Ways to Reduce W/C

The main rule that determines the concrete strength is the 'water-cement ratio rule'. According to the available data [27], for high and medium concrete classes, approximately 40% of the total strength is formed due to the adhesive bonding of cement-sand mortar with crushed stone; 20% due to mechanical adhesion caused by the crushed stone surface microrelief. For low concrete classes (for which the mortar component strength is less than 20 MPa), more than half of the total strength is contributed by adhesion. The growth of adhesion is achieved primarily by convergence of cement grains with aggregates, which contribute to reduction of W/C and water separation degree.

The value of the bonding strength decreases significantly with an increase in the aggregate size, which can be explained by the increase in the shrinkage effect as well as the processes of water separation and contraction in the concrete mixture. The concrete strength decreases with an increase in the content of crushed stone lamellar and needle-shaped grains from 25 to 50% or more, which is explained by the negative effect of the latter on the concrete mixture workability and, as a result, on its compacting.

Adhesion of solid bodies depends on the value of their surface energy, which is determined by the strength of crystal lattices as well as the properties of the environment that surrounds the body. It is possible to increase the activity of crushed stone and sand, using various technological methods. An increase in surface energy is achieved by mechano-chemical processing of materials, breaking of interatomic bonds. New, freshly formed aggregate surfaces obtained by grinding and other methods of mechanical processing have higher values of surface energy, which determines their increased adhesive activity. Following the available data, the use of freshly crushed stone increases the concrete compressive strength by up to 20%, tensile and flexural strength by up to 30% [28]. To obtain high-strength concrete, it is possible to re-crush crushed stone immediately before the concrete mixture production. It destroys the dusty-clay film on the grains and forms new surfaces, which help to increase the concrete strength.

Adsorption of vapour moisture and carbon dioxide from the air by crushed aggregates and saturation of uncompensated molecular forces leads to 'aging' of their surface and serves as a certain obstacle to formation of reliable adhesive contacts. In this regard, it is effective to create on aggregate grains a primary contact layer of structured binder. With this aim, quartz sand can be treated by lime. This increases the concrete flexural strength by up to 25%. Joint grinding of cement and sand also contributes to intensification of the structure formation processes at the contact zones. One of the ways to activate aggregates is to create an optimal relief of their surface. Increasing the roughness of the aggregate contributes to the mechanical jamming of the binder and also increases the contact surface area.

Cement stone adhesion with aggregates is significantly weakened by clay minerals and iron hydroxide films. Presence of these impurities in fine fractions leads to an increase in concrete mixtures' water demand, which also negatively affects the concrete strength. The treatment aggregates by acidic or alkaline solutions contributes both to creation of a developed aggregates microrelief and simultaneously cleans their surface. In addition, treatment of aggregates by certain solutions leads to a change in their surface charge, which also affects their reactive ability.

Activation of aggregates' adhesive ability due to an increase in their free surface energy can be achieved by the influence of electric and magnetic fields and by ultrasonic treatment. Under the action of an external electric field, molecules and ions, that make up solid bodies, are polarised, which contributes to an increase in adhesive strength. Adhesion of particles is also facilitated by a decrease in the wetting angle under the influence of an electric charge. Following available data [29], there is a positive effect of electric spark discharge during crushed stone crushing on its surface chemical activity. The flexural strength of specimens on quartzite and limestone rocks processed by electro-hydraulic crushing increased by more than 20%.

In modern concrete technology, plasticizers are the most widely used admixtures for reducing *W/C*. Reducing water demand and accordingly *W/C* without worsening the concrete mixtures' workability allows improving a number of concrete properties – strength, stability, impermeability, etc. By the plasticizing action effectiveness, i.e. increasing the concrete mixture workability without reducing the concrete strength, plasticizers are divided into four categories (Table 2.4). Unlike ordinary plasticizers, which reduce water demand to 10-15%, superplasticizers (SP) allow reduction by 20-30% or more and, accordingly, increase in the concrete strength.

Table 2.4: Classification of plasticizers for concrete mixtures

Category	Name	Increase of Sl from 20-40 mm	Decrease in water demand, %
I	Superplasticizers	to 200 mm and more	At least 20
II	Plasticizers	140-190 mm	At least 10
III	Plasticizers	90-130 mm	At least 5
IV	Plasticizers	<80 mm	<5

Superplasticizers began to be used in concrete production in the early 1970s. It resulted in significant improvement of concrete properties without increasing the cement consumption, to obtain cast and self-compacting concrete mixtures with a moderate water demand, to produce high-strength concrete using ordinary Portland cement and aggregates with low permeability, high corrosion resistance, etc. Addition of SP is currently a mandatory condition in producing high-quality, high-tech HPC.

Classification of superplasticizers (Table 2.5) divides them by composition and mechanism of action. The SP action is caused by a complex of physical and chemical processes in cement paste-admixture system. The SP mechanism of action is determined mainly by:

1. Adsorption of mono- or poly-molecular surfactants on the surface of mainly hydrated new formations.
2. Colloid-chemical phenomena at the boundary of phases separation.

Table 2.5: Classification of superplasticizers

Type	SP composition	Mechanism of action	Relative cost of the polymer, %
NF	Based on sulfonated naphthalene formaldehyde polycondensates	Electrostatic	40
MF	Based on sulfonated melaminoformaldehyde polycondensates	Electrostatic	80
LST	Based on desugared lignosulfonates	Electrostatic	20
P	Based on polycarboxylates and polyacrylates	Steric	100

The SP action of NF, MF, LST types (Table 2.5) is dominated by the effect of cement particles' electrostatic repulsion due to the fact that SP molecules adsorption layers increase the zeta potential value on the cement particles' surface. The value of zeta potential, which has a negative sign, depends on the SP adsorption capacity. An increase in the SP adsorption capacity is facilitated by an increase in the hydrocarbon chain length and molecular weight. The SP adsorption is proportional to their concentration in an aqueous solution. Among cement minerals, C_3A has the highest adsorption capacity, β-C_2S has the lowest.

The electrokinetic potential of the cement particles' surface upon adding SP changes from $+11$ to $-(25\text{-}35)$ mV, which causes mutual repulsion of uniquely charged particles. In the mechanism of action SP of type P, the role of zeta potential is lower and the mutual repulsion of cement particles is ensured by the so-called steric effect. This effect is caused by the chains' shapes and the nature of the charges on the surface of cement grains and hydrates.

Widely used SP of the naphthalene formaldehyde type (NF) are produced both in the form of 20-40% solutions and powder. When the admixture is dosed at 0.5-1% of the cement weight, it allows to increase the concrete mixture slump from 20-40 mm to 200-220 mm. Under conditions of equal concrete mixtures' workability, concrete mixture produced with addition of such SP have lower W/C and the obtained compressive strength at 28 days is 30-50% higher than for concrete without SP. At the same time, the concrete density and water tightness significantly increase and a number of other concrete properties improve.

Using SP of the new generation type P based on polycarboxylates provides an increase in the concrete mixture slump from 30 mm to 210-250 mm and more at relatively low dosages. If concrete mixtures with addition of traditional SPs quickly lose their workability, mixtures with polycarboxylates SP remain in a plastic state for 1.5-2 hours. The high storage capacity of concrete mixtures with this type SP makes them especially attractive for monolithic construction and at long-term transportation. Like other SPs, they are successfully used at heat-moisture processing of concrete in precast concrete industry.

Along with effective plasticizers of new generation, complex admixtures, which, along with SPs, contain well-known and much cheaper admixtures with a lower plasticizing effect (for example, technical lignosulfonates, LST), are also of interest. Under certain conditions, and first of all, the use of intensive methods of compaction, reduction of W/C and obtaining HSC is possible with the use of stiff and ultra-stiff concrete mixtures. The highest effect in using stiff mixtures occurs when combining dynamic and static pressure in the compaction process [24].

Regulating the Hydration and Structure Formation Processes

This direction in the technology of high-strength concrete can be implemented by a set of measures, including the regulation of cement grinding fineness, adding accelerating and highly dispersed mineral admixtures that accelerate structure formation, increase curing temperature of concrete.

The most universal and effective way of modifying the concrete structure and regulating its properties is introduction of admixtures into the concrete mixture. Currently, in economically developed countries, almost all concrete is produced with the use of various admixtures. The types of known admixtures are extremely diverse. As a rule, admixtures have a multifunctional effect on concrete mixtures and hardened concrete. There are different classifications of admixtures according to the main effect of their action. Common classification suggests distinguishing four classes of chemical admixtures:

- Regulators of concrete mixtures' rheological properties, their ability to save over time;
- Regulators of concrete mixtures setting and hardening, kinetics of their heat generation;
- Regulators of concrete porosity, ensuring its corrosion resistance, frost resistance, waterproofing;
- Adding special properties to concrete (hydrophobic, electrically conductive, biocidal, etc.).

Another classification of admixtures according to the main technological effect is proposed by the ASTM [30] (Table 2.6).

Table 2.6: Classification of chemical admixtures according to ASTM

Type	Technological effect	Standard No.
A	Water demand decrease	C494
B	Retardening of hardening	C494
C	Acceleration of hardening	C494
D	Water demand decrease/retardening of hardening	C494
E	Water demand decrease/acceleration of hardening	C494
F	High decrease in water demand/retardening of hardening	C494
G	Plasticizing for cast concrete	C1017
	Plasticizing/retardening of concrete hardening	C1017
	Air entrainment	C260
	Admixtures for shotcrete	C1141

The group of hardening accelerators includes, as a rule, electrolyte salts, the main effect of which is accelerating of concrete mixtures' hardening (in some cases, setting time). Of the concrete hardening accelerators, calcium chloride has been studied to the maximum extent. This admixture was proposed in 1885 [30]. The effect of calcium chloride is explained by the increase in the cement clinker minerals' solubility, formation of complex sparingly soluble substances, catalytic and modifying effect during cement hydration. Its use in concrete, however, is limited due to the acceleration of reinforcement steel corrosion and a decrease in the cement stone stability in a sulphate environment. In some countries, the use of this admixture is prohibited.

Sodium and potassium sulphates, sodium and calcium nitrates, iron chloride, aluminium chloride and sulphate and other electrolyte salts are also used as hardening accelerators.

In recent years, the accelerating effect of sodium thiosulphate and rhodanide ($Na_2S_2O_3$ and NaSCH), which is similar to the effect of $CaCl_2$, on cement hardening has been established. The concrete strength at the early hardening stages increases proportionally to the concentration of these admixtures. Addition of sodium thiosulphate and rhodanide do not cause corrosion of reinforcement in reinforced concrete. Both thiosulphate and sodium rhodanide are relatively expensive admixtures, therefore mixtures of these salts based on industrial waste, in particular coke gas processing, are of practical interest.

Portland cement hardening acceleration is mainly caused by accelerating the alite phase hydration. Following Ramachandran [31], cations and anions are arranged in rows due to the accelerating effect on C_3S hydration:

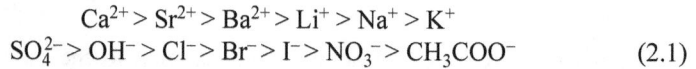

$$Ca^{2+} > Sr^{2+} > Ba^{2+} > Li^+ > Na^+ > K^+$$
$$SO_4^{2-} > OH^- > Cl^- > Br^- > I^- > NO_3^- > CH_3COO^- \qquad (2.1)$$

Chemical admixtures-accelerators are used to increase the early strength of concrete, reduce the cement consumption, reduce the products heat-processing duration, and the heating temperature. The effectiveness of using hardening accelerators is higher; the shorter the cycle of heat-moister processing, the lower the cement and concrete classes of strength and the isothermal heating temperature.

According to the effect on the cement stone strength immediately after steaming, the admixtures can be placed in the following order [32]:

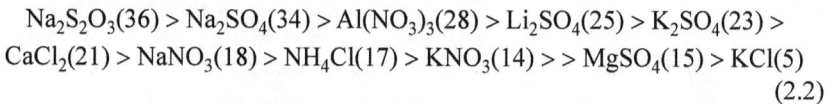

$$Na_2S_2O_3(36) > Na_2SO_4(34) > Al(NO_3)_3(28) > Li_2SO_4(25) > K_2SO_4(23) >$$
$$CaCl_2(21) > NaNO_3(18) > NH_4Cl(17) > KNO_3(14) > > MgSO_4(15) > KCl(5) \qquad (2.2)$$

In parentheses is given the cement stone strength increase in per cent, compared to the reference.

Added 28 days after steaming, the effectiveness of the admixtures is as follows:

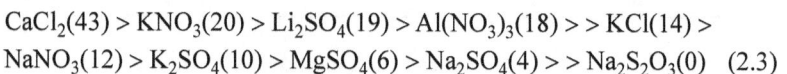

$$CaCl_2(43) > KNO_3(20) > Li_2SO_4(19) > Al(NO_3)_3(18) > > KCl(14) >$$
$$NaNO_3(12) > K_2SO_4(10) > MgSO_4(6) > Na_2SO_4(4) > > Na_2S_2O_3(0) \qquad (2.3)$$

Trying to universalise the action of admixtures and enhance their technical effect led to the wide spread of complex (composite) admixtures. Complex admixtures can be divided into two categories. The first is represented by mixtures of admixtures belonging to the same class, and the second by mixtures of admixtures belonging to different classes.

All complex polyfunctional modifiers (PFM) can be divided into four groups: I – mixtures of electrolytes; II – mixtures of surfactant; III – mixtures of electrolytes and surfactants; IV – mixtures of chemical and mineral admixtures.

Mineral admixtures are finely dispersed admixtures made of mineral materials, which are added into concrete mixtures in an amount usually >5% to improve or give concrete special properties. By origin, admixtures of this type are both natural and man-made. The classification of active mineral admixtures adopted in the USA and European countries is based on their activity, chemical and mineralogical composition (Table 2.7).

Table 2.7: Classification and characteristics of mineral admixtures

No.	Classification		Chemical and mineral composition	Characteristics
	Feature	Material		
1	Binding properties	Rapidly cooled metallurgical slags	Silicate glass containing CaO, MgO, Al_2O_3. Crystalline components in a small amount	Granules with 5-15% moisture. After drying grained to size < 45 μm. Specific surface area (S_a) $S_a =$ 350-500 m^2/kg (according to Blaine)
2	Binding and pozzolanic properties	High-calcium fly ash (CaO >10%)	Amorphous silica containing CaO, MgO, Al_2O_3. Crystalline components may be present in the form of SiO_2, C_3A. Man-made hydraulic admixtures	10-15% of particle size >45 μm, most of them have spherical shape. $S_a = 400$ m^2/kg (according to Blaine)
3	High pozzolanic activity	Microsilica	Microsilica of amorphous modification	Powder (spherical particle shape) $d =$ 0.1 μm. $S_a \approx 20$ m^2/kg (according to BET)
		Rice husk ash		Particles <45 μm with a developed cellular structure. $S_a \approx 60$ m^2/kg (according to BET)
4	Normal pozzolanic activity	Low-calcium fly ash (CaO <10%)	Silicate glass containing Al_2O_3, Fe_2O_3, alkalis. Crystalline substance from SiO_2, mullite, hematite, magnetite	Powder (spherical particle shape) >45 μm. Most of particles >20 μm. $S_a \approx 250$-350 m^2/kg (according to Blaine)
		Natural materials, firewood	In addition to aluminosilicate glass, contains quartz, feldspar, mica	Most particles ground to size of <45 μm, acute structure
5	Weak pozzolanic activity	Slowly cooled slags, hydroremoval ashes, ash slags	Crystalline silicate materials with few non-crystalline components	Additionally crushed to provide pozzolanic properties

The group of active admixtures or pozzolan includes materials capable of entering into a chemical reaction with calcium hydroxide at normal temperature to form substances with binding properties. During concrete hardening, the source of calcium hydroxide is the main minerals that are included in Portland cement clinker and undergo hydrolysis under the influence of water.

Highly active mineral admixtures for concrete that are increasingly widely used in recent decades, include ultra-dispersed waste from production of ferroalloys, the so-called microsilica (MS). MS is a condensed aerosol that is caught by filters of gas-cleaning systems at melting metallurgical furnaces. It contains spherical particles with an average diameter of 0.1 μm and a specific surface area of 15,000-25,000 m^2/g and more. Its bulk density is 150-250 kg/m^3. According to the chemical composition, MS is represented mainly by non-crystalline silica, the content of which usually exceeds 85 and reaches 98%. MS as an admixture for concrete was first proposed in the early 50s and began to be used in a large scale from the early 70s of the last century in Norway, and then in other countries. According to Norwegian standards, the amount of silicon dioxide in MS should be at least 85% and the dosage of the admixture in concrete should not be more than 10% of the cement weight. The unique specific surface in combination with the amorphised particles structure, the presence of such impurities as silicon carbide, which have a high surface energy, determine the high structuring ability and reactivity of this material in comparison with other active mineral admixtures.

Due to its extremely high dispersion and amorphous particles' structure, MS causes a significant increase in concrete mixtures' water demand; therefore it is used in combination with superplasticizers. Along with MS, other mineral materials can serve as effective modifiers of concrete under certain conditions (high dispersion, combination with superplasticizers, etc.) – metakaolin, zeolites, etc.

One of the ways to improve the concrete structure formation is to reduce capillary porosity and improve properties by impregnating concrete with polymeric substances or monomers followed by polymerisation (obtaining concrete polymers). When concrete is impregnated, its structure changes; first of all, the open capillary porosity is radically reduced, the contact zone of cement stone with aggregates is compacted. As a result, water absorption decreases and strength increases significantly; other physical and mechanical properties improve, wherein concrete of lower strength is characterised by a higher strengthening coefficient.

Porosity of dense concrete varies within the limits of 6-20%, the volume of pores with a diameter of more than 1 μm can reach 2-3%. Reducing the concrete porosity by 10% increases the concrete strength approximately twice. The features of the material porous structure determine the choice of impregnating substances and treatment modes. Less viscous substances can penetrate into thinner pores and capillaries, but with their help, it is difficult to ensure the monolithisation of big pores and defects. Polymers formed in concrete pores create a three-dimensional network that reinforces the silicate base. They also fill irregularities, pores and cracks on the aggregate surface, changing the structure of the cement stone contact zone with aggregate.

Cement Concrete with High Early Strength

For fast construction of structures by using general purposes construction cements, the problem of ensuring, along with high concrete strength at 28 days, a rather high early strength (at one day) is urgent. For HPC the strength at two days ranges from 30 to 50 MPa.

According to the theoretical ideas discussed above, ensuring early concrete strength is possible when reducing W/C to the minimal possible values with a simultaneous increase in cement hydration degree. The concrete strength on conditional aggregates is proportional to the cement stone strength. At equal cement stone strength, the concrete strength is higher; the greater the modulus of elasticity of coarse aggregate, the better its adhesion to cement stone. In cases when the aggregates' strength is not lower than that of the cement stone, at sufficient adhesion between them it has low significant effect on the concrete strength.

According to T. Powers [33], compressive strength of cement stone samples hardened under normal conditions; $f_{c.s}$ corresponds to the following equation:

$$f_{c.s} = AX^n \tag{3.1}$$

where A is a constant characterising the cement gel strength (A $\approx$ 240 MPa); n is a coefficient determined by the cement characteristics ($n = 2.6$-3); X is a structural criterion calculated according to Eq. (1.37).

The structural criterion X in Eq. (3.1) characterises the cement hydration products' concentration in the space available for these substances.

Structural criterion X, justified by Powers, unlike the criterion of R. Fere [1], which was first proposed for predicting the concrete strength in 1892, is a parameter proportional to the relative density of cement stone and not of the cement paste. Provided the cement hydration degree (α) is known, it allows predicting the cement stone strength at any age for a given W/C value. Dependencies close to Powers equation, which take into account the relationship between the cement stone strength and its relative density, were proposed later by other researchers [10]. Calculated values of cement stone strength, obtained according to the Powers equation for different values of W/C and α, are given in Table 3.1.

Table 3.1: Effect of W/C and α on cement stone strength

W/C	α	Cement stone strength following Eq. (3.1)	W/C	α	Cement stone strength following Eq. (3.1)
0.2	0.2	32.7	0.3	0.2	13.3
	0.3	73.8		0.3	32.7
	0.5	178.6		0.5	89.8
				0.7	160.1
0.25	0.2	20.1	0.35	0.2	9.3
	0.3	47.7		0.3	23.4
	0.5	124.0		0.5	67.3
				0.7	124.0

It follows that at extremely low W/C values, there are opportunities for increasing the cement stone strength even with a relatively small increase in cement hydration degree α. For example, the transition from $\alpha = 0.2$ to $\alpha = 0.3$ at $W/C = 0.2$ allows to bring strength of the cement stone $f_{c.s}$ to 73.8 MPa, while at $W/C = 0.3$, the estimated values $f_{c.s}$ at $\alpha = 0.3$ are only 32.7 MPa, i.e. more than twice lower.

Known experimental data [34] also demonstrates a higher increase in cement stone strength at low W/C values with a relatively small increase in hydration degree. For example, following F. Locher [35], at $W/C = 0.2$ with an increase in α from 0.1 to 0.2, the cement stone compressive strength increases from 30 to 55 MPa, and already at $W/C = 0.3$ – only from 15 to 25 MPa. This conclusion is of fundamental importance in developing the technology of high-strength, rapid-hardening concrete. At the same time, it should be borne in mind that the calculated dependence of type (1.37), relating the cement stone strength to its hydration degree when using cements with low water demand, can give significant deviations from experimental data. It is known [36], in particular, about the discrepancy between relatively low hydration degree of cements with low water demand and their high strength, which is caused by composition and structure of hydrated products based on these binders.

Increasing hydration degree at the early hardening stages of cement with a certain chemical and mineralogical composition is achieved by a complex of known technological methods and mostly by increasing the specific surface by increasing the finest particles (<5-10 microns) content at grinding as well as using hardening accelerating admixtures [37]. From the published experimental data [38] it follows that an increase in grinding fineness from 300 to 500 m²/kg, as well as use of a number of accelerating admixtures, most significantly increases the Portland cement hydration degree at early hardening stages after one to three days. This conclusion is explained [10, 37, 38] by the fact that approximately

24 hours after mixing, dense shielding shells from hydrates are formed on the cement grains and inhibit the further cement hydration process. At the same time, as a result of the crystallisation pressure, stresses occur in the cement stone, which slow down the growth of its strength in the following, after the initial period, hardening time [10]. With an increase in cement grinding fineness and optimal gypsum content for each level of dispersal, along with hydration degree, the cement strength continuously increases at one to three days and at 28 days, it increases only to certain limits for specific surface area of 410-520 m²/kg [38].

The effect of accelerating admixtures is also more significant in the early stages of cement stone hardening up to a certain optimal dosage [38]. Important conditions for positive effect of high cement grinding fineness and hardening accelerators on strength, along with achieving higher hydration degree and, as a result, lower capillary and general porosity, are the pore size reduction and improvement of the hardening cement stone structure [10; 37; 38]. The maximum hydration degree, which practically approaches a value of one during hardening in water, is possible for a cement paste W/C of at least 0.42. When W/C <0.42, the maximum value of $\alpha = 2.38\ W/C$ [10, 38].

Equations like (1.37) allow to predict the cement stone strength depending on W/C and α, as well as to solve the inverse problem of finding the specified parameters at a given strength of cement stone. The type of equations and the values of coefficients taken into account in them may change depending on the characteristics of the cement composition, its granulometry, hardening conditions, type and content of admixtures.

Table 3.2 presents the ultimate compressive strength values of alite Portland cement at its hardening in water and normal temperature [39]. These values were obtained by testing cubic samples with an edge length of 3.16 cm, made from

Table 3.2: The influence of cement stone hardening duration at normal hardening on the strength and hydration degree [39]

Indicators	Specimens hardening duration, days					
	1	*3*	*7*	*28*	*90*	*180*
Ultimate cement stone compressive strength, MPa ($f_{c.s}$) [39]	17.2	28.4	45.7	70.2	86.7	92.4
Experimental values of the hydration degree (α_e) [39]	0.2	0.31	0.35	0.43	0.51	0.60
Calculated values of the hydration degree (α_c) according to Eq. (1.38)	0.215	0.2713	0.350	0.467	0.546	0.575
α_c according to Eq. (1.37)	0.1810	0.2229	0.2820	0.3616	0.422	0.4230

Notes: 1. When calculating according to Eq. (3.1), the value of A is assumed to be 240; $n = 2.7$.

2. Below the line are given the relative deviations of the calculated values of α compared to the experimental (in %).

cement paste with $W/C = 0.25$. Table 3.2 also show experimental and calculated values of α.

The calculated values of α_c in Table 3.3 are obtained according to the following equation:

$$\alpha_c = \frac{f_{c.s} + K_1}{K_2} \tag{3.2}$$

Table 3.3: The effect of additional cement grinding and its type on the hydration degree (α_c) and strength of cement stone

No.	W/C	S_a of cement, m²/kg	Hydration degree (α)/Cement stone strength, MPa, at				
			12 h	1 day	2 days	7 days	28 days
			Portland Cement CEM I				
1	0.2	350	0.19 / 29.3	0.28 / 64.7	0.33 / 88.2	0.36 / 103.2	0.41 / 129.4
2	0.2	450	0.31 / 78.5	0.39 / 118.8	0.41 / 129.4	0.45 / 151	0.47 / 162
3	0.25	350	0.22 / 24.8	0.3 / 47.7	0.35 / 64.7	0.4 / 83.3	0.46 / 107.3
4	0.25	450	0.35 / 64.7	0.41 / 87.2	0.46 / 107.3	0.51 / 128.3	0.55 / 145.5
5	0.3	350	0.23 / 18.3	0.34 / 42.4	0.41 / 61.7	0.49 / 86.6	0.58 / 117
6	0.3	450	0.37 / 50.4	0.45 / 73.8	0.55 / 106.6	0.61 / 127.6	0.65 / 141.9
			Portland Cement CEM II/A				
1	0.2	350	0.18 / 29.0	0.26 / 63.2	0.31 / 86.8	0.34 / 101.1	0.40 / 127.5
2	0.2	450	0.30 / 78.2	0.38 / 117.3	0.39 / 128.8	0.42 / 151.5	0.46 / 160.8
3	0.25	350	0.20 / 21.6	0.27 / 44.1	0.31 / 60.9	0.35 / 78.4	0.41 / 100.1
4	0.25	450	0.34 / 62.4	0.39 / 86.6	0.45 / 104.9	0.48 / 126.1	0.52 / 145.9
5	0.3	350	0.24 / 16.1	0.35 / 40.7	0.41 / 57.6	0.48 / 83.5	0.56 / 112.3
6	0.3	450	0.36 / 48.0	0.45 / 71.5	0.56 / 105.9	0.60 / 125.1	0.66 / 138.1

where $f_{c.s}$ is the cement stone strength at n days; K_1 and K_2 are coefficients, which for the experimental data given in Table 6, correspond to $K_1 = 27.5$ and $K_2 = 210$, respectively.

When using Eq. (3.2), a high correlation of experimental and calculated values of α_c is achieved, which, obviously, can be explained by the fact that this equation is 'tied' to a specific cement and its hardening conditions.

Table 3.3 demonstrates experimental data on the influence of changes in grinding fineness and cement type on the cement stone's early strength at low W/C values. Portland cement types CEM I and CEM II/A with the following calculated mineralogical composition of clinker were used: C_3S – 68.4%, C_2S – 13.05%, C_3A – 7.05%, C_4AF – 11.5%. The initial specific surface area of cement was $S_a = 350$ m^2/kg. The cement compressive strength under normal hardening conditions was: at two days – 38.7 MPa, at seven days – 41.1 MPa, at 28 days – 53.1 MPa. To obtain the strength indicators, cubic specimens were prepared from cement paste, which hardened under normal conditions – 12 hours, one, seven and 28 days.

Analysis of the data presented in Table 3.3 confirms the above discussed patterns and shows the possibility of achieving at $W/C = 0.2$-0.25 and specific surface of cement $S_a = 450$ m^2/kg sufficiently high cement stone strength values not only at one day and later, but also at normal hardening even at 12 hours. At low W/C, the replacement of Type I Portland cement with Type II has practically no effect on the obtained values of hydration degree and cement stone strength.

Table 3.4 and Fig. 3.1 present the concrete strength values for mixtures of the same workability on Portland cement of Types I and II without and with addition of naphthalene-formaldehyde superplasticizer in an amount of 0.7% by cement weight. The use of the admixture at constant cement consumption increased the strength at 28 days by approximately one class and the strength at one day increased by 27-51%. Further studies were carried out, using a typical medium-

Table 3.4: Increase in the early compressive strength of concrete due to adding naphthalene-formaldehyde superplasticizer

Cement type	W/C = 0.57 (without superplasticizer)				W/C = 0.47 (with superplasticizer)			
	Sl, mm	Strength, MPa, at			Sl, mm	Strength, MPa, at		
		1 day	7 days	28 days		1 day	7 days	28 days
CEM I	105	8.8	15.8	28.2	120	12.4	22.7	35.7
CEM II/B	100	6.55	13.8	23.85	120	9.0	17.5	29.9
CEM II/A	150	7.5	13.95	26.0	160	11.3	20.2	33.5

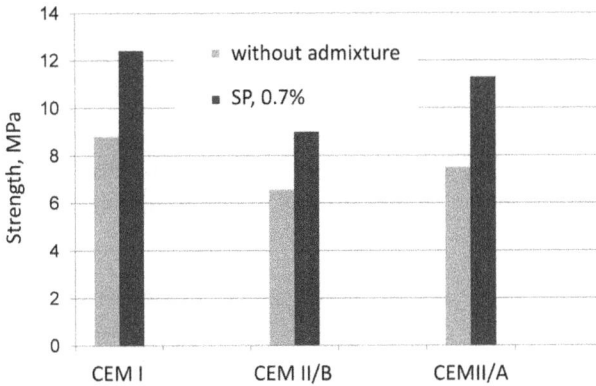

Fig. 3.1: Increase in concrete strength at $W/C = 0.47$ at one day of hardening due to decrease in concrete mixture water demand by using superplasticizer

aluminate Type I cement. Quartz sand with $M_f = 1.8$ and granite crushed stone fraction 5-20 mm were used as aggregates.

Table 3.5 compares the effectiveness of the investigated plasticizing admixtures for reducing water demand while maintaining the concrete mixture

Table 3.5: Comparative effectiveness of plasticizers

Type of plasticizing admixture	Consumption, % cement weight (for dry matter)	Water-reducing ability, %		Average increase in strength at one day, %	
		Cement paste	Concrete mixture	Cement stone	Concrete
Lignosulfonate	0.2	7-10	8-12	8-14	10-15
Sika Plastiment BV	0.3	9	10-12	9-18	10-20
Naftaline-formaldehide (S-3)	0.35	10-13	12-15	10-22	10-20
	0.5	15-16	16-18	16-31	15-30
	0.7	18-21	18-20	32-48	29-49
Polyacrylate and Polycarboxylate					
Mapei Dynamon SP3	0.2	28-31	30-35	40-58	42-60
	0.35	39-42	40-45	55-85	60-90
Mapei Dynamon SR3	1	20-25	22-28	40-55	38-52
	1.5	26-32	30-35	40-58	38-55
Melflux 2651F	0.5	30-32	30-35	42-63	40-60
	1	38-42	40-45	60-90	57-89

Note: The water-reducing effect was determined by the formula: $WRE = \dfrac{W - W_0}{W_0} \cdot 100\%$, where W_0 is water content without the addition of plasticizer, W – with plasticizer.

workability (Sl = 120-150 mm) and increasing the strength at one day. Admixtures of polycarboxylate type (Mapei Dynamon SR3, Melflux 2651F BASF) have the highest water-reducing ability, which allows to achieve the maximum increase in early strength.

As an alternative to increasing the cement specific surface, it is possible to consider using hardening acceleration admixtures. Table 3.6 presents the results on the influence of adding calcium nitrate (NC) on the hydration degree and strength of cement stone at reduced *W/C*.

Table 3.6: The influence of the water-cement ratio and hardening accelerating admixture on hydration degree and cement stone strength

No.	W/C	Hydration Degree (α)/Cement Stone Strength, MPa, at				
		12 h	1 day	2 days	7 days	28 days
Without Accelerating Admixture						
1	0.25	0.22	0.3	0.35	0.4	0.46
		24.8	47.7	64.7	83.3	107.3
2	0.3	0.23	0.34	0.41	0.49	0.58
		18.3	42.4	61.7	86.6	117
Calcium Nitrate (1.5% of cement weight)						
3	0.25	0.30	0.34	0.40	0.45	0.52
		64.5	73.5	81.4	89.8	115.5
4	0.3	0.35	0.41	0.46	0.52	0.61
		57.1	60.1	67.8	79.9	104.5

It should be noted that the kinetics of cement stone strength growth on finely ground alite cement and cement with hardening accelerator addition at *W/C* = 0.2-0.3 is significantly different from the traditional one and is characterised by an increase in compressive strength after 12 hours up to 50% and at one day, up to 70% of the value achieved at 28 days.

In order to study the effect of increasing the cement hydration degree due to accelerator admixtures in high-strength concrete with low *W/C* values on the kinetics of concrete strength growth, a complex effect of polycarboxylate type superplasticizers (Melflux) with various types of accelerating admixtures was studied.

Adding Melflux polycarboxylate type superplasticizer as the main admixture enabled to obtain a concrete mixture with slump 160-220 mm at *W/C* = 0.25. For specimens without the admixture, the compressive strength at 28 days was 102 MPa. After 12 hours of hardening the concrete compressive strength was 26-30 MPa, and at one day – 52.6 MPa, which is 30 and 50% of the strength at 28 days, respectively (Table 3.7).

Table 3.7: Complex effect of polycarboxylate type superplasticizer with various types of accelerating admixtures

Type and content of admixture	W/C	Slump, mm	Compressive strength (MPa)						
			12 h	1 day	2 days	7 days	28 days	90 days	180 days
Without admixture	0.48	150	7.2	16.6	23.2	35.5	50.2	57.2	58
Basic Admixture Melflux 2651F (0.5%)									
-		200	43.6	52.6	76.8	96	102	109.5	110.8
Relaxol*- Temp - 3, 1.5%		160	44.2	54.4	75.2	81.2	100.2	109.2	112
Relaxol- Antifreeze, 1.5%	0.25	220	56.2	72.4	82.8	102.8	106.8	108.9	109.6
Relaxol- Super, 1%		160	55.6	73.2	92.2	110	111.4	113	114.6
Relaxol- Universal, 7%		130	58.4	72	89.2	102.8	115.2	115.2	115.2
Calcium nitrate, 1.5%		210	48.4	70	83	94.2	114.2	120.5	121.5

* Relaxol – mixture of sodium thiosulfate and thiocyanate

Hardening-accelerating admixtures create an additional effect on strength at an early stage. Using 'Relaxol' admixtures of yields an increase in strength already from 12 hours of hardening: (up to 55-58 MPa). Tendencies of hardening acceleration at 12 hours are preserved at later hardening periods.

After 28 days of hardening, almost all tested concrete specimens showed an average increase in strength of 6-9% (up to 107-109 MPa).

According available data [40], the use of ultra-low *W/C* in concrete can lead to negative consequences related to durability of such concrete. Possibility of destructive phenomenon due to accelerated growth of the hydrated products volume in the limited inter-granular cement stone space was reported [40]. This process is associated with acceleration of hydration and a high initial concentration of clinker. Such phenomenon can cause decrease in high-strength concrete strength, especially at later periods. Therefore, it was of interest to trace the kinetics of strength growth for high-strength concrete at 90 and 180 days. The results of the research are shown in Table 7. According to the obtained data, long-term concrete hardening at *W/C* = 0.25 with various types of chemical admixtures occurred with a slight increase in strength in the range of 5-10%, due to the increased hydration degree in the initial period. No decrease in the high rapid hardening concrete strength was observed. Thus, the conducted studies confirm the theoretical justifications that in concrete at low *W/C* (of 0.25-0.3), produced by using effective superplasticizers, it is possible to ensure high strength concrete at significant increase in strength at the early stages (12 hours to one day) due to increase in the hydration degree by thin cement grinding and using hardening accelerating admixtures.

Experimental-statistical Models of Strength Parameters of High-strength Rapid-hardening Concrete (HSRHC)

Formulas of the general type for predicting the strength of concrete are convenient; however, under the influence of a significant number of factors, it is advisable to use experimental-statistical models of the relevant parameters. For this purpose, it is advisable to use the methodology of mathematic planning of experiments [41-43].

Mathematical planning of experiments is understood as setting up experiments according to a pre-compiled scheme with optimal properties. Mathematical planning makes it possible to solve the problem of building regression equations that relate the initial parameters with various controlled quantitative factors and use these equations as models for analysing their influence, technological calculations and optimisation with the minimum possible number of experiments.

The experiment is planned in accordance with a typical matrix and the permissible area of factors variation (factor space) is selected on the basis of a preliminary study of the object in accordance with the set goal.

For experimental studies of the influence of a set of given factors on the strength and other properties of concrete, it is usually advisable to use plans of the second order, in which the factors are planned at three levels.

The results of the experiments, after their statistical processing, make it possible to obtain the quadratic equations of the regression of the form

$$y_1 = b_0 + \sum_{i=1}^{k} b_i x_i + \sum_{i=1}^{k} b_{ii} x_i^2 + \sum_{i \neq j} b_{ij} x_i x_j + \dots$$

Obtaining the model can be considered completed after statistical analysis, which confirms their adequacy.

Given below are examples of obtaining and analysing experimental-statistical models of the strength of high-strength concrete under different initial conditions and a set of varied factors.

Task 1: *Obtain and analyse experimental-statistical models of the strength of HSRHC when using Portland cement CEM I with a possible compressive strength at 28 days of age in the range of 53-65 MPa, quartz sand with a fineness modulus of $M_f = 1.8$ and granite crushed stone of a fraction of 5-20 mm. The water consumption of the concrete mixture ensures its slump of 160-200 mm. Metflux – polycarboxylate type superplasticizer is introduced into the concrete mixture.*

The planning conditions of the experiment are given in Table 4.1, obtained during the implementation of plan B_3 [42] experimental-statistical models in Table 4.2.

Table 4.1: Conditions for planning the experiment

Factors		Levels of variation factors			Variation interval
Natural view	*Coded view*	*−1*	*0*	*+1*	
Water-cement ratio, *W/C*	X_1	0.25	0.3	0.35	0.05
Cement consumption, *C*, kg/m³	X_2	500	550	600	50
Cement strength of R_c, MPa	X_3	53	59	65	9

Table 4.2: Experimental-statistical models of strength of HSRHC

Regression Equation

$$f_{cm}^{2\,days} = 46.1 - 18.8X_1 - 0.3X_2 + $$
$$+ 6.4X_3 + 1.5X_1X_2 + 1.1X_1X_3 + 1.2X_2X_3 \qquad (4.1)$$

$$f_{cm}^{28\,days} = 99 - 14.7X_1 + X_2 + 3.6X_3 - 0.5X_1X_2 + $$
$$+ 0.4X_1X_3 - X_2X_3 \qquad (4.2)$$

The compressive strength of concrete was determined after 12 hours, one, two and 28 days.

The analysis of the obtained polynomial models of concrete strength (Figs. 4.1-4.3) showed a decisive influence *W/C* on the strength of concrete in the studied range, both at an early and at a later age, practical insignificance at constant *W/C* and cement strength R_c (X_3) cement consumption, which varies in a certain area. This conclusion is consistent with known data [39].

The set of obtained equations allows to use them both for predicting the strength of concrete at a given age and for designing the compositions of concrete mixtures.

In Eq. (4.1-4.2):

$$X_1 = \frac{W/C - 0.3}{0.05}; \quad X_2 = \frac{C - 550}{50}; \quad X_3 = \frac{R_c - 55}{5}$$

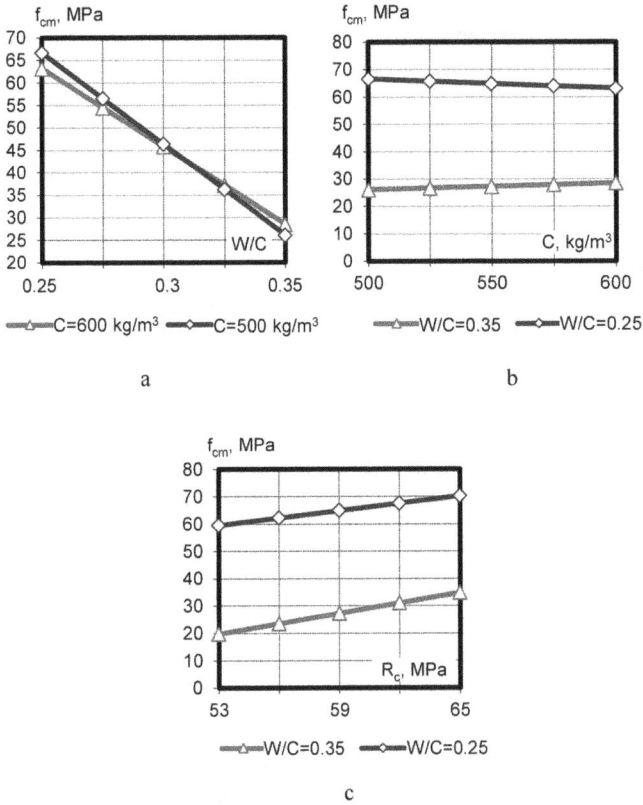

Fig. 4.1: Graphs of dependence of the compressive concrete strength at the age two days from: a, b – water-cement ratio and cement consumption; c – water-cement ratio and cement strength

where W/C, C and R_c are, respectively, water-cement ratio, cement consumption, kg/m³ and cement strength, MPa.

For practice in calculating the compositions of concrete with additives in a wide range of C/W (4.1-4.2), the most convenient formulas are:

$$f_{cm} = kAR_c(C/W - b) \qquad (4.3)$$

where k is a coefficient that depends on the type of admixture and its content.

According to our experimental data, when the polycarboxylate type of superplasticizer was introduced into concrete mixtures, the average K values for concrete at the age of two days were 1.7; 28 days – 1. An increase in the value of K when shortening the duration of hardening shows that the meaning of the strength of cement used increases at an early age, which confirms the thesis substantiated above about the special importance of increasing the initial degree of hydration of cement to ensure high early strength of cement stone.

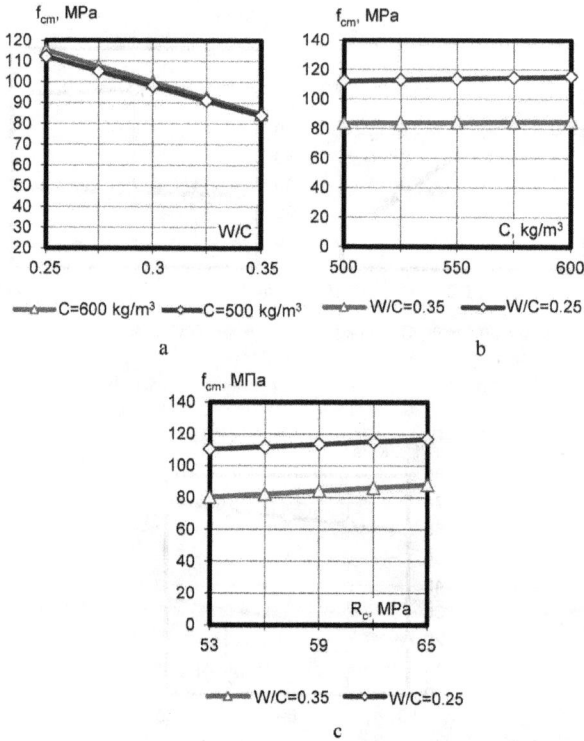

Fig. 4.2: Graphs of the dependence of compressive concrete strength at the age of 28 days from: a, b – water-cement ratio and cement consumption; c – water-cement ratio and cement strength

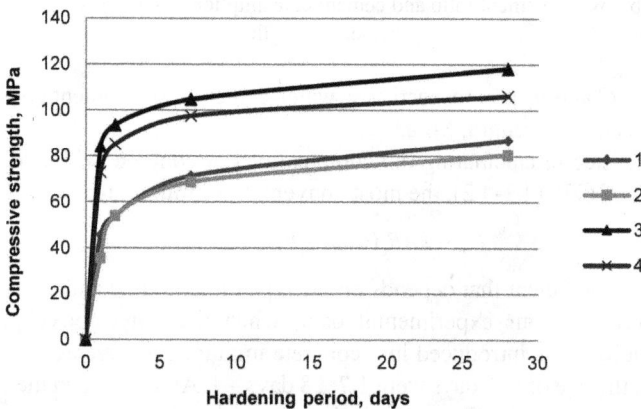

Fig. 4.3: Kinetics of a set compressive strength high strength rapid hardening concrete:
1 – $W/C = 0.35$, $R_c = 50$ MPa; 2 – $W/C = 0.35$, $R_c = 60$ MPa;
3 – $W/C = 0.25$, $R_c = 50$ MPa; 4 – $W/C = 0.25$, $R_c = 60$ MPa

For more accurate calculations of the required values of W/C or C/W for concrete that provides the required strength indicators at a certain age, it is advisable to take the values of the coefficients in formula (4.9) based on empirical data tied to specific materials or use special dependencies.

The equations of regression of the strength of HSRHC at different ages obtained by us transformed/translated into Eqs. (4.4-4.5) for the convenience of calculating the composition of concrete. Calculation equations of concrete strength are given below:

$$f_{cm}^{2\,days} = 0.59 R_c (C/W - 1.92); \tag{4.4}$$

$$f_{cm}^{28\,days} = 0.48 R_c (C/W + 0.22) \tag{4.5}$$

Equations 4.4-4.5 can be used to determine C/W, which ensures the concrete strength ratio f_{cm}^n / f_{cm}^{28} at the required age. Further calculation is carried out according to known methods [5].

Statistical processing of experimental data shows that the value of this ratio at a constant temperature can be satisfactorily described by equation 4.6 and taking into account the possible influence of temperature in the range 5-40°C (Eq. 4.7).

$$f_{cm}^n / f_{cm}^{28} = a \cdot \ln(n) + b \tag{4.6}$$

$$f_{cm}^n / f_{cm}^{28} = a_1 \cdot \ln(n) + b_t \cdot t + c \tag{4.7}$$

where n is the duration of hardening, days; t is the hardening temperature, °C.

The calculated values of the coefficients in Eqs. (4.6) and (4.7) are given in Table 4.3.

Table 4.3: Values of coefficients in equations 4.6 and 4.7

Design concrete strength, MPa	Eq. (4.6) t = 20°C		Eq. (4.7) t = 5-40°C		
	a	b	a_1	b_t	c
20-30	0.242	0.24	0.242	0.0115	0.008
35-50	0.218	0.33	0.218	0.0134	0.063
≥ 60	0.166	0.4473	0.166	0.0133	0.1806

An increase in the predictive ability of Eq. (4.3), when designing concrete compositions is achieved by introducing of the multiplicative coefficient $pA = A_1 \cdot A_2 \cdots A_n$, which takes into account a complex of factors affecting the strength of concrete, including the duration and temperature of hardening.

If it is necessary to predict the strength with a known composition of concrete or to find the necessary C/W to ensure the specified strength at a certain temperature and duration of hardening, the calculation equation takes the general form:

$$f_{cm} = (a \cdot \ln(n) + b_t \cdot t + c) \cdot [A \cdot R_u (C/W - b)] \tag{4.8}$$

Equation 4.8, specified for certain starting materials and a range of concrete strength classes, as well as a temperature interval, can also be used to calculate the necessary values of the duration of hardening and temperature to achieve the specified strength. Equations 4.7 and 4.8 with the coefficients given in Table 4.3 are valid for $n = 1$-28 days and $t = 5$-40°C (Fig. 4.4).

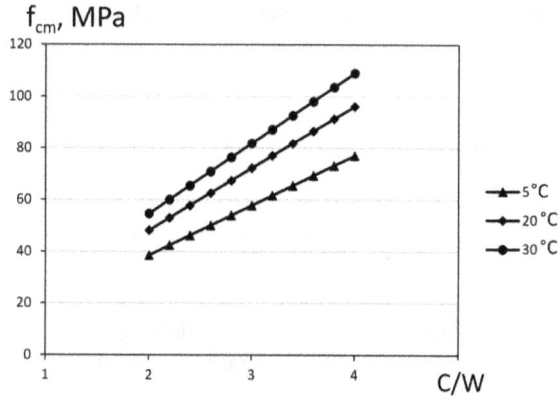

Fig. 4.4: Calculated dependencies of strength in 28 days from C/W
at different curing temperatures

The use of formulas of the type 4.3 for calculating the compositions of concrete subjected to heat-moisture treatment requires empirical classification of the multiplicative coefficient pA, taking into account the mode features of steaming. Such clarification can be performed, using the analysis of relevant experimental–statistical models. Table 4.4 shows the values of coefficients $A_{\tau,t}$ for high-strength concrete with possible changes in temperature ($X_1 = 40$-80°C) and duration ($X_2 = 4$-8 hours) of isothermal exposure during heat-moisture treatment.

Analysis of the formulas 4.9-4.12, given in Table 4 confirms the high energy efficiency of high-strength concrete. Already two hours after heat treatment, with a duration of isothermal exposure of four hours and a temperature of 80°C, their strength approaches 70% of the vintage. At a temperature of 40°C and a duration of isothermal heating of eight hours strength after steaming exceeds 50% of design. One day after curing at 60°C and four hours of isothermal exposure, the strength of concrete reaches 85% of the original strength.

The joint use of equations 4.5 and 4.9-4.12 makes it possible to solve a number of tasks of designing both the composition of concrete mixtures and the technological parameters of concrete production with a given strength, and in particular:

• to determine the necessary values of C/W and R_c at given heat treatment parameters;
• determine the necessary values of temperature and duration of heat treatment, which provide acceptable or minimum possible values of C/W;

Table 4.4: Values of coefficients $A_{\tau,t}$ for high-strength concrete

Compressive strength of steamed concrete, MPa	The value of coefficient $A_{\tau,t}$
2 hours	$A_{\tau,t}^{2\,hours} = 0.62 + 0.16x_1 + 0.12x_2 - 0.05x_1^2 - 0.04x_2^2 - 0.09x_1x_2$ (4.9)
1 day	$A_{\tau,t}^{1\,day} = 0.84 + 0.08x_1 + 0.04x_2 - 0.05x_1^2 - 0.01x_2^2 - 0.005x_1x_2$ (4.10)
7 days	$A_{\tau,t}^{7\,days} = 0.86 + 0.09x_1 + 0.04x_2 + 0.04x_1^2 - 0.004x_2^2 - 0.04x_1x_2$ (4.11)
28 days	$A_{\tau,t}^{28\,days} = 0.94 + 0.04x_1 + 0.02x_2 + 0.02x_1^2 + 0.003x_2^2 - 0.01x_1x_2$ (4.12)

Note: In equations (4.17) to (4.20), $X_1 = (t_{is}\text{-}60)/20$, $X_2 = (\tau_{is}\text{-}6)/2$, where t_{is} and τ_{is} are, respectively, the temperature and duration of isothermal heating of concrete during steaming

- to determine the required duration of hardening of concrete after heat treatment, which allows, with the given parameters of the composition, to achieve the required yield strength or to adjust these parameters, taking into account the possible duration of exposure of the products before their shipment.

Example 1: *Determine the necessary C/W values for obtaining concrete with a compressive strength at the age of 28 days $f_{cm}^{28\,days} \geq 100$ MPa with the achievement of strength after two days ($f_{cm}^{2\,days}$) ≥ 50 MPa. Take $R_c^{28\,days} = 60$ MPa.*

Tentatively, with the help of formulas 4.4, 4.5 at $A = 0.65$ and the values k given above, we consistently find the value of C/W and establish the entire set of necessary properties at $R_c = 60$ MPa.

To ensure the required concrete strength after two days:

$$C/W_1 = \frac{50}{0.59 \cdot 60} + 1.92 = 3.33;$$

- after 28 days:

$$C/W_2 = \frac{100}{0.48 \cdot 60} - 0.22 = 3.25.$$

The entire set of specified strength indicators of concrete is provided when using formula 4.4 at $C/W = 3.33$.

Example 2: *To determine the necessary value of C/W to obtain concrete with compressive strength at the age of 28 days $f_{cm}^{28} \geq 80 M\Pi a$ with the achievement of 50% of the design strength after two days of hardening. Compressive cement strength in 28 days $R_c = 52.5$ MPa. Calculate the minimum temperature value at which the specified strength index can be reached at the age of two days. Calculate also how much it is possible to shorten the duration of concrete hardening to* *achieve a strength of 80 MPa when the hardening temperature is increased to 40°C.*

Preliminarily, with the help of formula 4.5, we find the value of C/W, which will ensure the achievement of the required concrete strength at the age of 28 days

$$\frac{C}{W} = \frac{f_{cm}^{28}}{0.73 R_c} + 0.33 = \frac{80}{0.73 \cdot 52.5} + 0.33 = 2.42$$

According to Eq. 4.6, taking into account the coefficients given in Table 4.3, we will calculate the corresponding value of C/W to ensure the required strength of 40 MPa at the age of two days:

$$\frac{C}{W} = \frac{40}{(0.166 \cdot \ln(2) + 0.0133 \cdot 20 + 0.1806) \cdot 0.73 \cdot 52.5} + 0.33 = 2.19$$

According to the given calculation, the entire set of given concrete strength indicators is provided at $C/W = 2.42$.

Calculate with the help of equations the minimum value of the concrete hardening temperature, at which the necessary value of the strength of the concrete at the age of two days will be provided at $C/W = 2.42$.

$$t = \frac{\dfrac{f_{cm}^{2days}}{f_{cm}^{28}} - a\ln(n) - c}{b_t} = \frac{\dfrac{40}{80} - 0.166 \cdot \ln(2) - 0.1806}{0.0133} = 15.3°C$$

To ensure the required strength of 40 MPa on the second day, the minimum temperature of concrete hardening should be at least 15.3°C. However, the value of the strength of concrete at the age of 28 days at such a hardening temperature, as follows from Eq. 4.8, will be:

$$f_{cm}^{28} = (0.166 \cdot \ln(28) + 0.0133 \cdot 15.3 + 0.1806) \cdot 0.73 \cdot 52.5 \cdot$$
$$\cdot (2.42 - 0.33) = 75.1 \ MPa$$

According to the formula 4.8, we calculate the necessary duration of concrete hardening until it reaches a strength of 80 MPa when the hardening temperature is increased to 40°C.

$$\ln(n) = \frac{\dfrac{f_{cm}^{28}}{f_{cm}^{28}} - b_t \cdot t - c}{a} = \frac{1 - 0.0133 \cdot 40 - 0.1806}{0.166} = 1.73$$

Accordingly, $n \approx 2$ days.

Example 3: *Determine the necessary value of C/W to obtain concrete with compressive strength 28 days after heat-moisture treatment (HMT) $f_{cm}^{28} \geq 100 M\Pi a$, reaching after two hours of heat-moisture treatment (duration*

of isothermal exposure of eight hours, temperature of isothermal exposure 60°C)
70% of the design strength ($f_{cm}^{HMT} = 70\,M\Pi a$). Take $R_c^{28} = 52\,MPa$.

Preliminarily, using formulas 4.9 and 4.10, we find the value of the coefficient $A_{\tau,t}$, which corresponds to the given heat treatment mode. To do this, we transform the values of temperature and duration of isothermal exposure into coded form:

$$X_1 = (60-60)/20 = 0,\, X_2 = (8-6)/2 = 1$$

$$A_{\tau,t}^{2h} = 0.62 + 0.16 \cdot 0 + 0.12 \cdot 1 - 0.05 \cdot (0)^2 - 0.04 \cdot 1^2$$
$$-0.09 \cdot 0 \cdot 1 = 0.7$$

$$A_{\tau,t}^{28} = 0.94 + 0.04 \cdot 0 + 0.02 \cdot 1 + 0.02 \cdot (0)^2 + 0.003 \cdot 1^2$$
$$-0.01 \cdot 0 \cdot 1 = 0.963$$

Using Eq. 4.5 with additional consideration of the calculated values of the coefficients $A_{\tau,t}^{2h}$ and $A_{\tau,t}^{28}$ we determine the necessary value of C/W, which will ensure the achievement of the specified strength of concrete after heat-moisture treatment ($f_{cm}^{HMT} = 7$ MPa) processing and 28 days of hardening ($f_{cm}^{28} \geq 100$ MPa).

$$C/W = (f_{cm}^{HMT})/(0.73 A_{\tau,t}^{2h}\, R_c) + 0.33 = (70/0.73 \cdot 0.7 \cdot 52) + 0.33 = 2.94$$

$$(C/W)_1 = (f_{cm}^{28}/0.73 A_{\tau,t}^{28}\, R_c) + 0.33 = (100/0.73 \cdot 0.963 \cdot 52) + 0.33 = 3.04$$

According to the given calculation, the entire set of given concrete strength indicators is ensured at $C/W = 3.04$.

Example 4: *Calculate whether the required release strength of concrete will be provided two hours after heat-moist treatment of at least 30 MPa at the value of C/W determined in the previous example (C/W = 3.04), but with a reduction in the duration of isothermal exposure to four hours, temperatures up to 40°C.*

Preliminarily, using formula 4.9, we find the value of the coefficient $A_{\tau,t}$, which corresponds to the specified HMT mode. To do this, we transform the values of temperature and duration of isothermal exposure into coded form: $X_1 = (40-60)/20 = -1$, $X_2 = (4-6)/2 = -1$.

$$A_{\tau,t}^{2h} = 0.62 + 0.16 \cdot (-1) + 0.12 \cdot (-1) - 0.05 \cdot (-1)^2$$
$$-0.04 \cdot (-1)^2 - 0.09 \cdot (-1) \cdot (-1) = 0.16$$

Using formula 4.5 and taking into account the calculated value of the coefficient $A_{\tau,t}^{2h}$ and the specified value of C/W, determine the strength of concrete of two hours after heat-moist treatment:

$$f_{cm}^{2h} = 0.73 A_{\tau,t}^{HMT} R_c (C/W - 0.33) = 0.73 \cdot 0.16 \cdot 52.5(3.04 - 0.33) = 16.6 \text{ MPa}$$

The given condition is not fulfilled. To ensure the required strength of concrete, it is necessary to apply a set of appropriate technological measures (adjustment of the HMT regime, introduction of admixtures-accelerators, etc.).

Task 2: *To obtain experimental–statistical models of the compressive and tensile strength of HSRHC at splitting, the composition of Portland cement was adopted: clinker – 50%, blast furnace slag – 12%, fly ash – 38%. To intensify the grinding of cement, a 0.04% by mass propylene glycol additive was introduced into the ball mill.*

As a coarse aggregate, granite crushed stone with a fraction of 5-20 mm was introduced into the concrete mixture, and quartz sand with $M_f = 1.95$ was used as a fine aggregate. As a plasticizing additive, a superplasticizer of the polycarboxylate type was used.

Chemical composition of blast furnace slag and fly ash is given in Tables 4.5 and 4.6.

Table 4.5: Chemical composition of blast furnace slag

Name of the indicator	Marking indicator	Quantitative value, %
Silicon oxide	SiO_2	39.1-39.9
Aluminium oxide	Al_2O_3	6.33-6.65
Iron oxide (III)	Fe_2O_3	0.11-0.18
Calcium oxide	CaO	46.8-47.4
Magnesium oxide	MgO	3.08-3.22
Manganese oxide	MnO	1.14-1.21
Sulphur oxide	SO_3	1.69-1.76

Table 4.6: Chemical composition of fly ash

Sample	$SiO_2+Al_2O_3+Fe_2O_3$, %	SO_3, %	Free CaO, %	MgO, %	$Na_2O + K_2O$, %	L.O.I., %
I	82.5	2.5	2.8	2.1	1.2	5.1
II	84.5	2.1	2.5	2.0	1.1	4.9
III	81.3	2.4	3.0	2.0	1.3	5.4
Average	82.8	2.3	2.8	2.0	1.2	5.1

The main studies of the influence of dispersion of composition cement (*CC*), the content of additives *SP* and *W/C* on the strength of concrete, were performed, using the three-level three-factor plan B_3 [42], the planning conditions are given in Table 4.7.

Table 4.7: Conditions for planning experiments

Factors		Factor variation levels			Interval variation
Natural view	Coded view	−1	0	+1	
W/C	X_1	0.25	0.3	0.35	0.05
Content of SP, %	X_2	0.4	0.7	1	0.3
Specific surface area, S_a, m²/kg	X_3	350	450	550	100

In the course of research, standard cube samples (10×10 cm) were made at each point of the plan to assess the influence of factors on the strength of concrete based on *CC*, which hardened under normal conditions. The compressive strength of cube samples at the age of one, seven and 28 days and the tensile strength at axial splitting in the age of one and 28 days were determined.

After processing and statistical analysis of experimental data, mathematical models of strength in the form of polynomial regression equations were obtained.

Statistical models of the strength of concrete based on *CC*:

$$f_{cm}^1 = 36.72 - 2.31X_1 + 1.68X_2 + 6.3X_3 + 0.36X_1X_2 + 0.72X_1X_3 -$$
$$0.21X_2X_3 + 0.21X_1^2 - 4.8X_2^2 - 0.91X_3^2 \tag{4.13}$$

$$f_{cm}^{28} = 80.1 - 6.09X_1 - 0.3X_2 + 9.57X_3 + 0.01X_1X_2 +$$
$$1.86X_1X_3 + 0.38X_2X_3 - 2.5X_1^2 - 4.28X_2^2 - 3.78X_3^2 \tag{4.14}$$

$$f_{c,tn}^{28} = 5.08 - 0.262X_1 - 0.013X_2 + 0.415X_3 + 0.069X_1X_3 +$$
$$0.001X_2X_3 + 0.027X_1^2 - 0.185 \cdot X_2^2 - 0.172X_3^2 \tag{4.15}$$

When analysing the mathematical models 4.13-4.15, a significant interaction of the factors *W/C* and S_a on both early and design strength is observed, while other interactions have a minor impact. A significant non-linear effect of the influence of all factors is observed at a later age.

The ranking of the quantitative effects of the influence of the studied factors on compressive strength indicators allows to place them in the following order: $X_3 > X_1 > X_+$. The influence of factors on the tensile strength at splitting has a similar character.

Graphic dependencies of concrete compressive strength at the age of one and 28 days, built on the basis of the obtained models, are shown in Fig. 5, for tensile strength during splitting in Fig. 6.

Analysing the obtained data and graphic dependencies (Fig. 5), we come to the conclusion that the compressive strength of the studied concretes at the age of one day lies within 20-43 MPa, at the age of 28 days – 59-92 MPa, at the consumption binder 500 kg/m³ (cement clinker content – 250 kg/m³ concrete mixture).

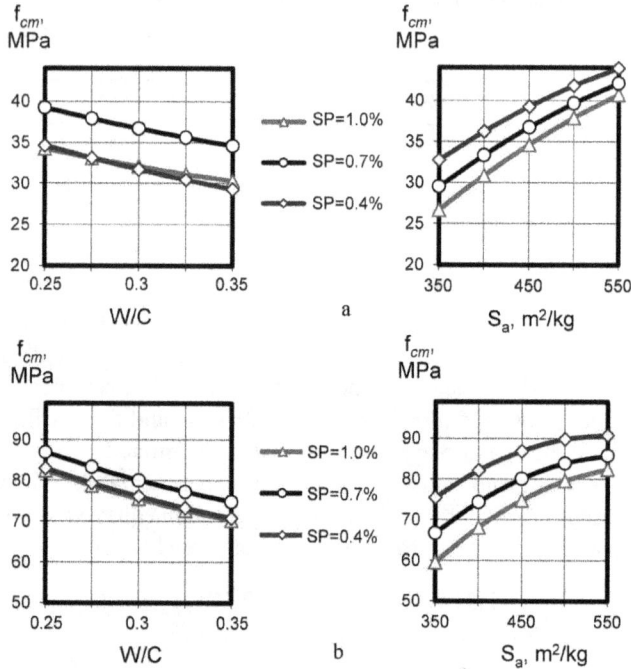

Fig. 4.5: Dependencies of concrete compressive strength
at the age of one (a) and 28 days (b)

An increase in the specific surface area of the CC from 350 to 550 m²/kg leads to a natural increase in strength by 35-50% during all hardening periods. However, at the age of one day, concrete on more fine cement (S_a = 450-550 m²/kg) have higher strength by 40-50% compared to S_a = 350 m²/kg, at a later age, the influence of dispersity is smoothed out. W/C also has a significant effect on both early and design strength. Increasing it from 0.25 to 0.35 causes a decrease in strength by 10-20 MPa.

According to the graphic dependencies (Fig. 4.6), the tensile strength of the tested concretes when split at the age of 28 days is 4.11-5.49 MPa, while the influence of the factors is similar to the influence on the compressive strength.

A nomogram (Fig. 4.7) was constructed on the basis of the obtained data and the mathematical model (4.14).

An important technological process that affects the structure and operational properties of concrete and reinforced concrete structures is heat moisture treatment (HMT).

The aim of the study was to establish the optimal parameters of the HMT regime, as well as its influence on the kinetics of strength gain and operational properties of concrete based on CC. A three-level four-factor plan B_4 [42] was implemented and the planning conditions are given in the Table 4.8. In the course

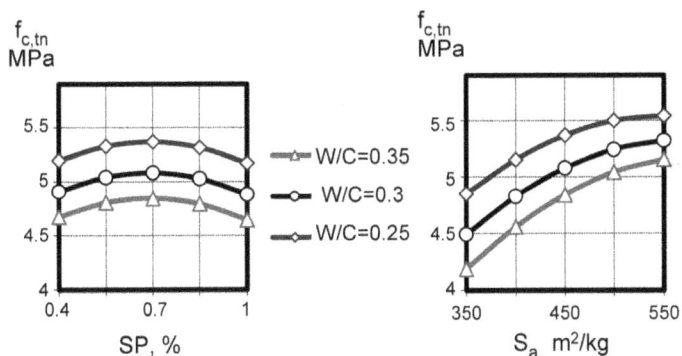

Fig. 4.6: Dependencies of concrete tensile strength on splitting at the age of 28 days

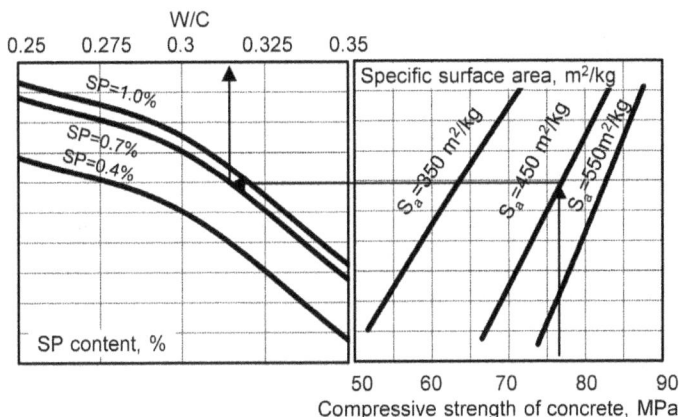

Fig. 4.7: Nomogram for determining the strength of concretes
based on *CC* at the age of 28 days

of research, standard cube samples ($10 \times 1 \times 10$ cm) were made at each point of the plan to assess the influence of factors on the strength of concrete, which hardened both during steaming and under normal conditions.

The water consumption of the concrete mixture was the same at all points. Heat treatment was carried out in a laboratory steaming chamber according to the following regime: pre-exposure – two hours; rising temperature with speed 25°C/h; isothermal aging – according to planning conditions (Table 4.8); the compressive strength of the samples was determined four hours after steaming and at the age of 28 days, as well as of the samples that hardened under normal conditions.

After processing and statistical analysis of the experimental data, the regression equation for the strength of steamed concrete was obtained:

• four hours after HMT:

$$f_{cm}^{HMT} = 58.3 - 0.46X_1 - 11.24X_2 + 1.36X_3 + 2.6X_4 + 0.8X_1X_2$$
$$+ 0.14X_1X_3 + 0.2X_2X_3 - 1.7X_1^2 + 4.0X_2^2 + 0.8X_3^2 + 1.85X_4^2; \quad (4.16)$$

- at the age of 28 days:

$$f_{cm}^{HMT\,28} = 65.04 + 0.46X_1 - 10.5X_2 + 0.5X_3 + 0.97X_4$$
$$- 0.49X_1X_2 + 0.24X_1X_4 - 0.3X_2X_3 - 0.16X_2X_4 \quad (4.17)$$
$$- 0.5X_1^2 + 4.2X_2^2 + 0.6X_3^2 - 0.6X_4^2$$

Table 4.8: Conditions for planning experiments when determining the optimal parameters of the HMT mode

Factors of influence		Levels of factor variation			Variation interval
Natural type	Coded type	−1	0	+1	
The content of SP additive in the *CC*, %	X_1	0.4	0.7	1.0	0.3
Water-cement ratio, *W/C*	X_2	0.25	0.3	0.35	0.05
Time of isothermal exposure, τ_{is}, hours	X_3	4	6	8	2
Maximum temperature, T_{is}, °C	X_4	60	75	90	15

Graphical dependencies of the strength of steamed concrete based on ash-containing *CC* on technological factors four hours after HMT are shown in Fig. 4.8, and at the age of 28 days in Fig. 4.9.

Their analysis makes it possible to place the factors affecting the strength of concrete after HMT in the following order of importance: $X_2 > X_4 > X_3 > X_1$. As expected, the *W/C* factor is the most influential, but some of its interaction with heat treatment parameters – duration and maximum temperature – can be traced. Among these two factors, temperature is more influential – an increase in the duration of isothermal exposure by two hours allows to compensate for a decrease in the maximum temperature by 15°C, at other equal concrete steaming conditions.

When increasing the isothermal exposure from four to eight hours, we observe an increase in strength within the range of 10-15%, as well as when increasing the temperature of HMT by 25-30°C. It should be taken into account that an increase in the temperature of HMT increases energy consumption more significantly than an increase in duration, moreover, 'soft' modes are necessary for increasing the durability of products. An increase in the amount of superplasticizer in the binder from 0.4% to 0.7-1.0% may be accompanied by a slight decrease in the strength of steamed concrete, but in this case, mixtures of a cast consistency with a spreading cone of up to 60 cm are obtained, which is a positive point in the production of modern high-tech concrete capable of self-compaction.

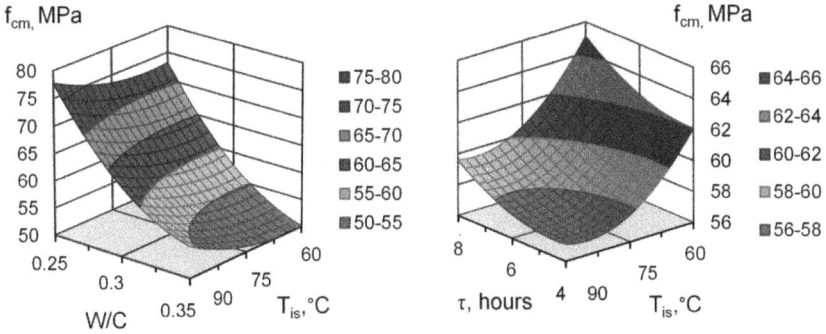

Fig. 4.8: The influence of technological factors on the strength of concrete after steaming

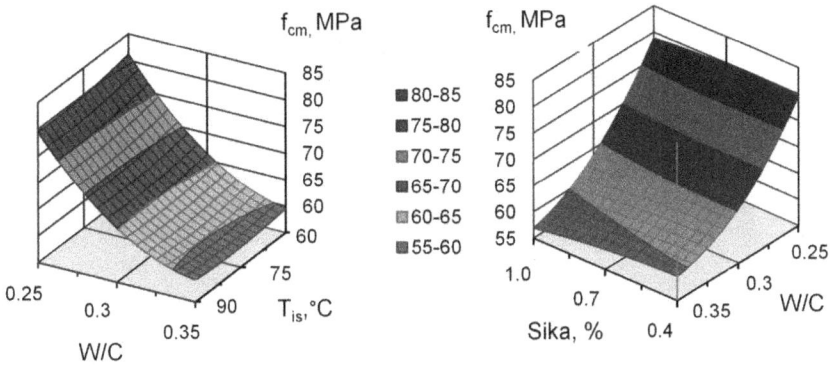

Fig. 4.9: The influence of technological factors on the strength of steamed concrete at the age of 28 days based on ash-containing *CC*

At the age of 28 days (Fig. 4.9), the strength of concrete with different HMT regimes practically equalises. In all cases, the strength of normal hardening samples is lower than that of steamed samples even at low temperature regimes of HMT. This can be explained by a more significant intensification of the hardening processes during HMT of concretes based on ash-containing *CC* than concretes based on ordinary Portland cement.

Example 5: *Find the necessary W/C to obtain concrete based on composite cement with a compressive strength at the age of one day (f_{cm}^1) at least 40 MPa, 28 days (f_{cm}^{28}) – 80 MPa with a tensile strength at splitting at 28 days at least 5 MPa.*

To solve this problem, it is necessary to solve a system of three inequalities based on equations 4.13-4.15:

$$\begin{cases} 40 \leq 36.72 - 2.31X_1 + 1.68X_2 + 6.3X_3 + 0.36X_1X_2 + \\ 0.72X_1X_3 - 0.21X_2X_3 + 0.21X_1^2 - 4.8X_2^2 - 0.91X_3^2 \\ 80 \leq 80.1 - 6.09X_1 - 0.3X_2 + 9.57X_3 + 0.01X_1X_2 + \\ 1.86X_1X_3 + 0.38X_2X_3 - 2.5X_1^2 - 4.28X_2^2 - 3.78X_3^2 \\ 5 \leq 5.08 - 0.262X_1 - 0.013X_2 + 0.415X_3 + 0.069X_1X_3 + \\ 0.001X_2X_3 + 0.027X_1^2 - 0.185X_2^2 - 0.172X_3^2 \end{cases}$$

As a result of solving the system, the required value of X_1 (*W/C* according to the Table 4.7) should be determined by changing the values of X_2 and X_3, while the optimality criterion should be selected. Analysing the factors of mathematical models, we establish that it is advisable to choose the minimum value of factor X_3 (specific surface area of composite cement) as an optimality criterion. During cement production, increasing the specific surface is achieved by increasing the intensity of grinding, which is a very energy-intensive and technically difficult task.

It is expedient to solve such a system of inequalities by means of sequential sorting of values in the environment of electronic spreadsheets. Using the MS Excel 'Solver' add-on, the following variable values were obtained that correspond to the established conditions:

$$X_1 = -1;$$
$$X_2 = 0.14;$$
$$X_3 = 0.12.$$

At the same time, the calculated values of the initial parameters were: f_{cm}^1 = 40 MPa, f_{cm}^{28} = 84.5 MPa, $f_{c,tn}^{28}$ = 5.4 MPa. Coded values correspond to the following factor values: *W/C* = 0.25; *SP* = 0.74%; S_a = 462 cm²/h.

Example 6: *Find the necessary value of the specific surface area (S_a) of the composite cement, which allows obtaining with a superplasticizer content of 0.7% and W/C = 0.3 strength after one day not lower than 35 MPa, 28 days – 70 MPa with tensile strength at splitting in 28 days not lower than 4.5 MPa.*

We convert the given factor values into coded form:

$$W/C - X_1 = (0.3 - 0.3)/0.05 = 0, \quad SP - X_2 = (0.7 - 0.7)/0.3 = 0.$$

Substitute the given values into the equation 4.13-4.15

$$35 = 36.72 + 6.3X_3 - 0.91X_3^2 \tag{4.18}$$

$$70 = 80.1 + 9.57X_3 - 3.78X_3^2 \tag{4.19}$$

$$4.5 = 5.08 + 0.415X_3 - 0.172X_3^2 \tag{4.20}$$

After solving quadratic equations, we get:

According to the equation 4.18, $X_3 = -0.26$

According to the equation 4.19, $X_3 = -0.8$

According to the equation 4.20, $X_3 = -0.99$.

We choose the maximum value of X_3 ($X_3 = -0.26$) as it will satisfy all the conditions of the problem.

We convert the coded value of the factor into its natural form according to Table 4.7:

$$S_a = 450 + (-0.26) \times 100 = 424 \text{ cm}^2/\text{g}$$

Task 3: *To obtain experimental and statistical models reflecting the influence of aggregate grain composition on water consumption and strength of high-strength concrete.*

Works on the design of the grain composition of concrete aggregates [5] were aimed at ensuring the minimum voidness of mixtures of grains of different shapes and sizes. Two approaches have been developed to ensure dense mixtures of aggregate grains: the choice of discontinuous and continuous grain composition.

The method of dense mixtures receiving, used in the selection of the composition of asphalt concrete aggregates, is proposed by N.N. Ivanov. He adopted an ideal ratio of the volume of each subsequent fraction to the volume of the previous one, the maximum size of which is two times larger, equal to $k = 0.81$. When calculating according to Fuller's formula, such a ratio (assuming that the densities of each fraction are equal) is $k = 0.707$, according to Hummel's formula, when $n = 0.3$ $k = 0.812$. Fairly dense mixtures, according to N.N. Ivanov, can be obtained with a value of the coefficient k in the range of 0.65-0.8 [5].

If we take the content in % of the first fraction equal to a, then the content of the second will be $a_1 k$, the third $a_2 k$, etc. The number of the last fraction should be equal to $a_{n-1} k$.

The sum of the volumes of all fractions can be written in the following form:

$$a(1 + \kappa + \kappa^2 + \cdots + \kappa^{n-1}) = 100\% \qquad (4.21)$$

Therefore, the content of the first fraction (that is, the partial residue on the corresponding sieve):

$$a_1 = \frac{1-k}{1-k^n} \cdot 100 \qquad (4.22)$$

However, for cement concrete, the use of curves of dense mixtures of aggregates is rational only in some cases, for example, for pressed or vibro-pressed concrete made from particularly stiff mixtures.

Experimental studies of the influence of the grain composition of stone crushing screening and superplasticizer on the properties of concrete from fine-grained mixtures were carried out. The experiment was conducted using the mathematical plan 'mixture-technology-property' [42], which makes it possible to simultaneously vary the content of the main fractions of the aggregate and the parameters of the composition of the concrete mixture (cement consumption and chemical admixture content). Granite sand, Portland cement CEM I 42.5 and superplasticizer Mapei Dynamon SP-3 were used in the experiments.

The sand was divided into three main fractions: 2.5-10 mm, 0.63-2.5 mm and 0-0.63 mm. The conditions for planning the experiment are given in Table 4.9. According to the experiment plan, a concrete mixture with slump corresponding to the S4 class – 160-210 mm was prepared. The water-cement ratio (W/C) was adopted as the characteristic of the water consumption of the concrete mixture, which ensured the given workability. Cube samples of 10×10×10 cm were made from the concrete mixture, hardened under normal conditions and tested at the age of 28 days with determination of compressive strength (f_{cm}, MPa). Additionally, the specific surface area (S_a, cm²/g) of the aggregate was monitored.

Table 4.9: Conditions for planning the experiment

Factors		Levels variation		
Natural	*Coded*	*0*	*0.5*	*1*
Aggregate Mixture				
Content of fine crushed stone (2.5-10 mm), %	V_1	25	40	55
Content of coarse sand (0.63-2.5 mm), %	V_2	25	40	55
The content of fine sand fraction (0-0.63 mm), %	V_3	20	35	50
Technology				
		−1	*0*	*+1*
Content of superplasticizer (SP, %)	X_1	0	0.5	1
Cement consumption (C, kg/m³)	X_2	300	450	600

As a result of the experiment (Table 4.10), adequate mathematical models of the initial parameters were obtained, which are given below:

$$W/C = 0.46V_1 + 0.54V_2 + 0.52V_3 - 0.33V_1V_2 - 0.14V_1V_3$$
$$-0.27V_2V_3 - 0.21V_1X_1 - 0.09V_1X_2 - 0.26V_2X_1 - 0.07V_2X_2 \qquad (4.23)$$
$$-0.21V_3X_1 - 0.12V_3X_2 + 0.04X_1^2 + 0.09X_2^2$$

$$f_{cm} = 49.3V_1 + 34V_2 + 45.9V_3 - 2.7V_1V_2$$
$$-7.5V_1V_3 - 20.4V_2V_3 + 21V_1X_1 + 10.1V_1X_2$$
$$+19.5V_2X_1 + 9.23V_2X_2 + 21.6V_3X_1 + 10.1V_3X_2 \qquad (4.24)$$
$$+7.62X_1X_2 + 2.14X_1^2 - 5.5X_2^2$$

Table 4.10: Planning matrix and experiment results

No.	Coded values of factors					Output parameters		
	V_1	V_2	V_3	X_1	X_2	W/C	Compressive strength, MPa	Specific surface of the aggregate, cm^2/g
							f_{cm}	S_a
1.	1	0	0	-1	-1	0.92	21.5	6.87
2.	1	0	0	+1	+1	0.29	84.2	6.87
3.	0	1	0	-1	-1	1.00	19.8	7.49
4.	0	1	0	+1	-1	0.48	41.9	7.49
5.	0	1	0	+1	+1	0.31	77.1	7.49
6.	0	1	0	-1	+1	0.87	21.4	7.49
7.	0	0	1	-1	-1	1.03	17.4	4.60
8.	0	0	1	+1	0	0.33	69.8	4.60
9.	0	1	1	-1	+1	0.68	24.4	4.60
10.	0.5	0.5	0	0	-1	0.60	30.1	7.18
11.	0.8	0.2	0	-1	+1	0.66	27.2	6.99
12.	0.3	0	0.7	+1	+1	0.30	79.2	2.28
13.	0.5	0	0.5	+1	-1	0.44	47.7	10.73
14.	0.6	0	0.4	0	0	0.45	46.1	9.96
15.	0	0.4	0.6	0	-1	0.66	25.3	11.75
16.	0	0.5	0.5	-1	0	0.74	20.7	11.04

The water-cement ratio of the concrete mixture before reaching the specified slump (*W/C*) varied in a wide range from 0.22 to 1.0 (Figs. 4.10 and 4.11). This range is caused by the joint action of various influential factors: A change in factor X_2 (cement consumption) causes a decrease in *W/C* on an average from 0.6 to 0.45, factor X_1 (content of superplasticizer) from 0.85 to 0.3. The factors of the grain composition mainly change the water consumption in accordance with the change in the specific surface of the aggregate. An increase in the content of the 2.5-10 mm fraction reduces the *W/C* by 0.03-0.05, an increase in the 0.63-2.5 and 0-0.63 mm fractions mainly increases the water consumption of the concrete mixture. Along with this, the influence of the 0-0.63 mm fraction is ambiguous – when the content of the fraction increases to 40% of the varied range, the water consumption decreases and only further increases are observed. As is known [44], mineral fillers are able to increase the slump of mortars and concretes in the absence of influence on water consumption. In this case, a small amount of the fine fraction, which contains up to 40% of particles smaller than 0.16 mm, increases the fluidity of the mixture while levelling the negative impact due to the superplasticizer. With a further increase in the content of granite sand, the water

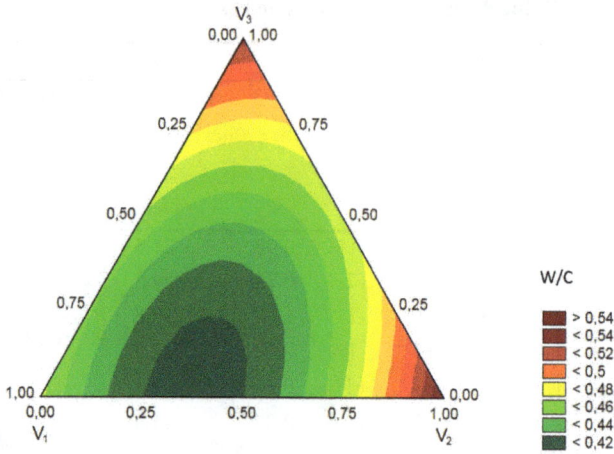

Fig. 4.10: Isoparametric diagram of the aggregate grain composition on water-cement ratio of high-strength fine-grained concrete (The values of the factors V_1-V_3 are given in Table 4.9)

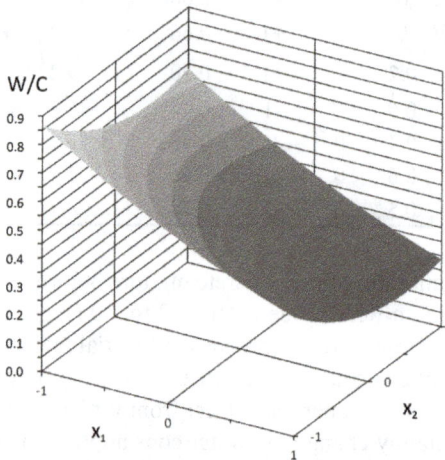

Fig. 4.11: Response surface of the water-cement ratio of fine-grained concrete

demand still increases. Comparing the effect of the aggregate grain composition on the water demand of high-strength fine-grained concrete with its effect on the specific surface area of the aggregate, it can be argued that the water demand value mainly correlates with fineness. However, with increased superplasticizer consumption, the influence of the grain composition becomes less noticeable.

The compressive strength of concrete at the age of 28 days varies from 15 to 82 MPa (Fig. 4.12). A significant interaction of factors X_1 and X_2 – superplasticizer content and cement consumption – is observed. The effect of grain composition on strength is basically consistent with the effect of components on water consumption (Figs. 4.12-4.13): the area of maximum strength of 75-82 MPa practically coincides with the minimum W/C (2.5-10 mm – 45-55%; 0.63-2.5 mm – 25-40%; 0-0.63 mm – 20-35%). As the superplasticizer content increases, the negative impact of dispersed particles on concrete strength decreases.

The obtained mathematical models make it possible to design the composition of high-strength concrete taking into account the grain composition of the aggregate, as well as to solve the inverse problem, in which it is possible to determine the necessary ratio of fractions of the aggregate to ensure the required strength of concrete.

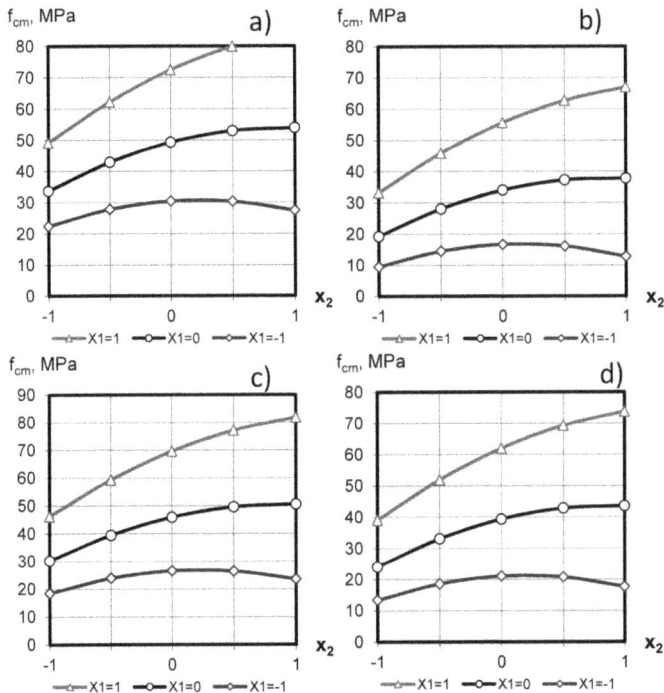

Fig. 4.12: Graphs of the influence of factors X_1 and X_2 (superplasticizer content and cement consumption) on the strength of fine-grained concrete with different fractions content in the aggregate (Table 9) (a) $V_1 = 1$; $V_2 = 0$; $V_3 = 0$, (b) $V_1 = 0$; $V_2 = 1$; $V_3 = 0$, (c) $V_1 = 0$; $V_2 = 0$; $V_3 = 1$, (d) $V_1 = 0.33$; $V_2 = 0.33$; $V_3 = 0.33$

Example 7: *Determine the composition of fine-grained concrete of class C45/55 (compressive strength of cubes at the age of 28 days – 70.7 MPa) with slump of the concrete mixture of 16-21 mm when using the following materials: Portland cement CEM I 42.5, granite sand fraction 0-10 mm (the grain composition of the sand is given in Table 4.11), superplasticizer of the polycarboxylate type Mapei Dynamon SP-3. The true density of cement is 3.1 g/cm^3, sand is 2.65 g/cm^3.*

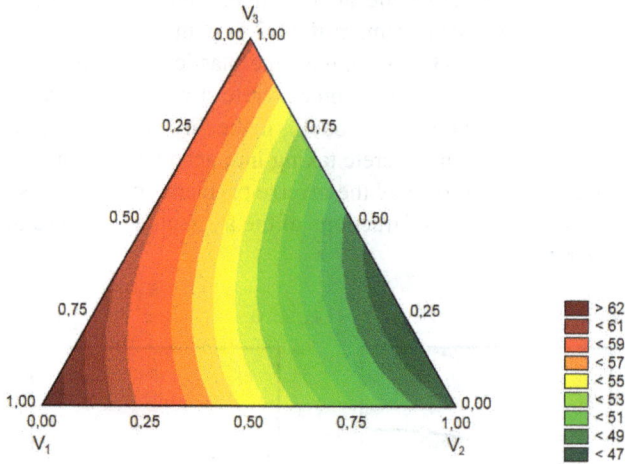

Fig. 4.13: Ternary graph of the response of the strength of fine-grained concrete (f_{cm}, MPa) to a change in the grain composition of the aggregate (when $X_1 = 0$ and $X_2 = 0$)

Table 4.11: Grain composition of granite sand, %

Partial Residues on Sieves with Cell Size, mm						
5	2.5	1.25	0.63	0.315	0.16	<0.16
7.2	34.7	10.5	14.2	13.6	10.9	8.9

1. Find the content of the main fractions in the aggregate in natural and coded values, necessary for the application of equations (4.23, 4.24) using the data in Table 4.11:

   - 2.5-10 mm – 41.9% (0.563)
   - 0.63-2.5 mm – 24.7% (0)
   - 0-0.63 mm – 33.4% (0.447)

2. Substitute the value of fraction content into equation 4.24:

$$f_{cm} = 49.3 \cdot (0.563) + 34 \cdot (0) + 45.9 \cdot (0.447) - 2.7 \cdot (0.563) \cdot (0)$$
$$-7.5 \cdot (0.563) \cdot (0.447) - 20.4 \cdot (0) \cdot (0.447) + 21 \cdot (0.563)X_1$$
$$+10.1 \cdot (0.563)X_2 + 19.5 \cdot (0)X_1 + 9.23 \cdot (0)X_2 \qquad (4.25)$$
$$+21.6 \cdot (0.447)X_1 + 10.1 \cdot (0.447)X_2 + 7.62X_1X_2 + 2.14X_1^2 - 5.5X_2^2$$

3. Solve the obtained equation with respect to X_1 and X_2, under the condition that the value of the X_1 factor (cement consumption) should be minimal. It is convenient to solve such an equation graphically. To do this, we build a diagram (Fig. 4.14) according to Eq. 4.24. The values that satisfy the given conditions are the following:

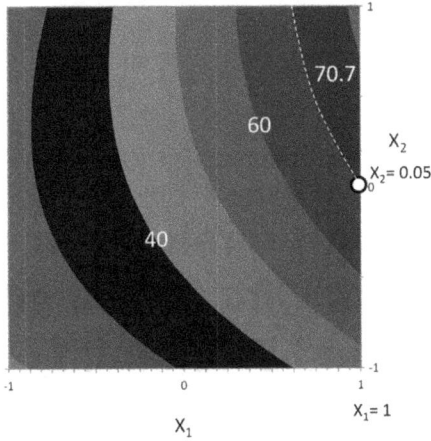

Fig. 4.14: Isoparametric diagram of the dependence of the strength of fine-grained concrete (f_{cm}, MPa) on factors X_1 and X_2, provided that the grain composition meets the conditions of the problem

$$X_1 = 0.05; X_2 = 1$$

4. Find the values of factors X_1 and X_2 in natural units:
   Consumption of cement $C = 450 + (0.05) \cdot 150 = 458$ kg
   Content of superplasticizer $SP = 1\%$
5. Substituting the found values of the factors into equation 4.23, we find the corresponding value of W/C:

$$W/C = 0.46 \cdot (0.563) + 0.54 \cdot (0) + 0.52 \cdot (0.447) - 0.33 \cdot (0.563) \cdot (0) -$$
$$-0.14 \cdot (0.563) \cdot (0.447) - 0.27 \cdot (0) \cdot (0.447) - 0.21 \cdot (0.563) \cdot (0.05) -$$
$$-0.09 \cdot (0.563) \cdot (1) - 0.26 \cdot (0) \cdot (0.05) - 0.07 \cdot (0) \cdot (1) -$$
$$-0.21 \cdot (0.447) \cdot (0.05) - 0.12 \cdot (0.447) \cdot (1) + 0.04 \cdot (0.05)^2 + 0.09 \cdot (1)^2$$

$W/C = 0.24$.

6. Find the water consumption: $W = C \cdot (W/C) = 458 \cdot 0.24 = 110$ l/m³.
7. Find the consumption of granite sand:

$$S = \left(1000 - \frac{C}{\rho_C} - W\right) \cdot \rho_S = \left(1000 - \frac{458}{3.1} - 110\right) \cdot 2.65 = 1967 \text{ кг/m}^3$$

8. Find the consumption of superplasticizer:

$$SP = 458 \cdot 1/100 = 4.6 \text{ kg/m}^3$$

Designing High-strength Concrete Compositions with Specified Values of Strength and Workability

The most developed and implemented in practice are two-parameter problems, when the normalised property of concrete is its compressive strength (f_{cm}) and the concrete mixture is an indicator of workability (Sl or Vebe). To solve problems of this type, computational and experimental methods are widely used which apply a number of known technological dependencies based on the rules of the water-cement (cement-water) ratio, stability of water demand of concrete mixtures, optimal sand content, etc.

When solving such problems for normal-weight concrete, the value of the cement-water or water-cement ratio (Fig. 5.1) is consistently determined as also the water consumption taking into account the necessary workability of the concrete mixture and consumption of aggregates, using the assumption that the concrete mixture consists of the absolute volumes of all its components. In the simplest case, for a four-component mixture, knowledge of three parameters is necessary: cement-water ratio (C/W), water consumption (W) and a factor characterising the ratio of aggregates (the proportion of sand in the mixture of aggregates (r) or the grains moving apart coefficient (K_a)). The last factor can be considered as an optimising one because only at some optimal value of it, on condition of C/W = const., it is possible to achieve the minimum consumption of cement. Most often, the ratio of aggregates is taken as optimal, which ensures the best workability of concrete mixture or minimal water consumption.

Multi-parameter problems arise when designing compositions of different types of concrete (hydraulic, road, corrosion-resistant, etc.). They can be divided into three subgroups:

- with normalised parameters uniquely related to the compressive strength of concrete;

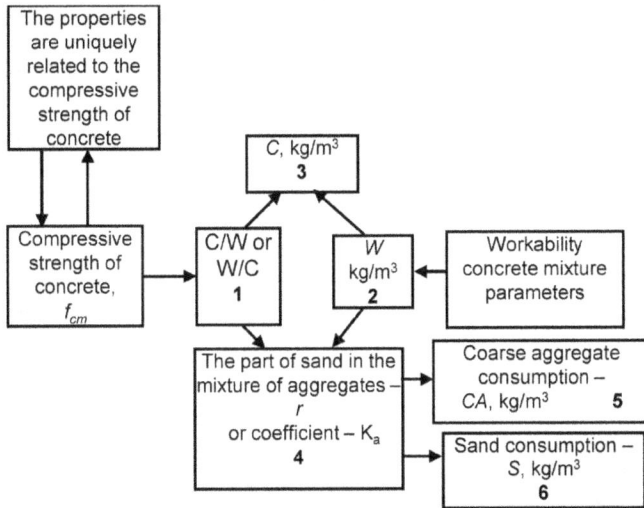

Fig. 5.1: Scheme of calculation of nominal compositions of normal-weight concrete

- with normalised parameters ambiguously related to compressive strength;
- with normalised parameters unrelated to compressive strength.

The first subgroup includes, for example, problems with different normalised concrete strength indicators. When calculating the composition of such concretes, first of all, as the determining parameter is found from the normalised properties of the concrete, its corresponding compressive strength and the minimum possible C/W is established, which ensures the entire set of properties. 'Determining parameter' means such a normalised parameter, the achievement of which ensures the achievement of all other parameters specified in the condition of the problem at the same time.

A fundamental feature of such problems is the existence of the C/W area, within which the C/W is located and which provides all the normalised indicators. The narrower this area is the closer the composition is to the optimal and $C \to$ min. To achieve this condition, various technological methods can be used: introduction of the admixtures-regulators of properties, change of hardening conditions, selection of starting materials, etc.

Mathematical planning of experiments (MPE) [42] as a method of mathematical modelling allows solving the design tasks of concrete and mortar compositions with a variety of initial conditions and factors.

The main advantages of mathematical modelling when solving the problems of designing concrete and mortar compositions are:

- the possibility of obtaining adequate under certain conditions of quantitative dependencies of indicators of normalised properties or their recipe-technological parameters (water demand, ratio of aggregates, water-cement

ratio, volume of entrained air, etc.), taking into account the influence of specific factors and the effects of their interaction;
- the possibility of calculating the composition of multi-component concrete and mortar mixes when normalising several initial parameters;
- the possibility of performing optimisation calculations and finding optimal compositions under given conditions and restrictions.

Mathematical models obtained by using MPE make it possible to solve the design tasks of concrete and mortar compositions together with the selection of mode parameters for various technological operations and to evaluate alternative solutions with the help of system analysis. For this purpose, the implementation of various algorithms, the use of analytical, grapho-analytical and graphic methods is possible.

Along with the advantages of mathematical models obtained by using MPE, for the design of concrete and mortar compositions, it should be taken into account that such models have a local character, i.e. they are valid under certain initial conditions and the use of specific materials in a certain area of varying factors. When changing the given conditions, the model should be used carefully and taken into account is the possibility of a significant increase in the error for the resulting solving. If it is necessary to estimate changes in factors that are not directly taken into account in the models, it is expedient to periodically adjust the coefficients of the models with the help of special adaptive algorithms.

Solving the problems of design of concrete and mortar compositions using complex mathematical models is expedient to perform with the help of computer programmes that allow to calculate basic compositions and correct them taking into account production information, statistical control of indicators of normalised properties.

Multifactor polynomial models make it possible to find the optimal values of such factors as the ratio of aggregates, the content of admixtures, etc., and thus to optimise the calculated compositions of concrete (mortar), taking into account the given set of factors and the range of their variation when determining the required compositions. At the same time, two approaches are possible:

- optimised factors are determined from equations where they play the role of dependent variables, for example, the proportion of sand in the aggregate mixture is found from the equation of the concrete mixture slump;
- obtain separate equations in which the optimised factors serve as output parameters and, together with the equations of normalised parameters, they are used in the calculation of concrete mixture compositions.

Below are examples of solving the problems of designing high-strength concrete compositions using the obtained experimental-statistical models.

Example 1: *Calculate the composition of high-strength rapid-hardening concrete with compressive strength at the age of 28 days $f_{cm}^{28\,days} \geq 80$ MPa with the achievement of compressive strength after two days $- f_{cm}^{2\,days} \geq 50$ MPa. The* *compressive strength of cement should be $R_c^{28\ days}$ = 60 MPa, its density ρ_c = 3.1 kg/l. The density of quartz sand is ρ_s=2.65 kg/l, its fineness modulus is equal to M_f = 2, the density of granite crushed stone is ρ_{cs} = 2.68 kg/l. The slump of the concrete mixture is Sl = 100-150 mm. The largest size of aggregate is 20 mm, bulk density ρ_b = 1.48 kg/l. Admixture of the polycarboxylate superplasticizer Melflux 2651F (0.5% of cement mass) is used.*

Preliminarily, with the help of formulas (4.4-4.5) at A = 0.6, we consistently find the value of C/W and establish the entire set of necessary properties at R_c = 60 MPa.

To ensure the required concrete strength after two days:

$$\left(\frac{C}{W}\right)_1 = \frac{50}{0.59\cdot 60}+1.92=3.33;$$

• after 28 days:

$$\left(\frac{C}{W}\right)_2 = \frac{80}{0.48\cdot 60}-0.22=2.55$$

The values of C/B required to ensure the specified strength indicators of high-strength concrete at different ages, found when using the proposed formulas (4.4-4.5), are quite close. For further calculations, we choose the maximum C/W = 3.33, which will provide the entire set of given concrete strength indicators.

The estimated water consumption to ensure the specified slump of 100-150 mm at a maximum aggregate size of 20 mm (in accordance with Table 5.1) will be:

$$W_0 = 220 \text{ l/m}^3$$

Taking into account the correction coefficients for the water-reducing effect when using the polycarboxylate-based plasticizing admixture Melflux 2651F K = 0.63 (Table 5.2), which is introduced in the amount of 0.5% of the cement mass, the total water consumption is:

$$W = W_0 \times K = 220 \times 0.63 = 139 \text{ l/m}^3$$

Then the consumption of cement is found by the formula:

$$C = C/W \times W = 3.33 \times 139 = 463 \text{ kg/m}^3$$

Find the consumption of crushed stone (CS) and sand (S) according to equations 4.5 and 4.6 with the coefficient K_a = 1.375 (Table 5.3) and inter-granular voids V_{cs} = 1 − $\rho_{b.cs}/\rho_{cs}$ = 1−1.48/2.68 = 0.45:

$$CS = \frac{1000}{\dfrac{1}{\rho_{cs}}+\dfrac{K_a V_{cs}}{\rho_{b.cs}}} \tag{5.1}$$

Table 5.1: Estimated water consumption (W_0) depends on the type of aggregates and ease of laying of the concrete mixture

Slump, mm	Water Consumption, l/m^3 at the maximum aggregate size, mm			
	10	20	40	70
160-200	237	228	213	202
120-160	230	220	207	195
100-120	225	215	200	190
80-100	215	205	190	185
50-70	210	200	185	180
20-40	200	190	175	170

Note: 1. When the content of silt and dust in crushed stone increases by more than 1% and particles smaller than 5 mm are above 5%, the water consumption increases by 1-2 litres for each percentage; When the content of silt and dust in the sand increases by more than 3% – by 2 l/m^3 for each percentage. 2. Consumption of water is given for concrete mixes made on crushed stone from igneous rocks. For concrete on gravel, water consumption is reduced by 10 l/m^3. 3. Water consumption is given for concrete mixtures based on Portland cement with a normal cement paste consistency of 26-28% and medium-grained sand (M_f = 2-2.5) without plasticizing mixtures. When changing the normal consistency of the cement paste for each percentage decreases, the water consumption decreases 3-5 litres, and increases by 3-5 litres. When changing the sand by module of fineness for every 0.5 in the downward direction, the water consumption increases by 3-5 l, in the upward direction, it decreases by 3-5 l.

Table 5.2: Correction factors when using plasticizing admixtures

Slump, mm	Cement-water ratio				
	1.4	1.8	2.2	2.6	3.0
-	0.96	0.95	0.94	0.93	0.92
	0.88	0.85	0.83	0.81	0.80
	0.77	0.75	0.73	0.71	0.7
10-40	0.93	0.92	0.92	0.92	0.91
	0.86	0.84	0.82	0.80	0.79
	0.76	0.74	0.72	0.7	0.69
50-90	0.91	0.91	0.90	0.90	0.89
	0.82	0.80	0.79	0.78	0.77
	0.71	0.7	0.69	0.68	0.67
100-160	0.90	0.89	0.88	0.87	0.87
	0.80	0.78	0.77	0.76	0.75
	0.69	0.67	0.66	0.65	0.64

Note: The upper row shows the values when using admixtures based on lignosulfonates in the amount of 0.25% of the cement mass, in the middle row – on the naphthalene-sulfonate basis in the amount of 0.7% of the cement mass, in the lower row – on a polycarboxylate basis – 0.7% of the weight of cement

Table 5.3: Grains moving apart coefficient (K_a)

Cement consumption, kg/m³	The value of K_a at W/C					
	0.3	*0.4*	*0.5*	*0.6*	*0.7*	*0.8*
250	–	–	–	1.26	1.32	1.38
300	–	–	1.3	1.36	1.42	–
350	–	1.32	1.38	1.44	–	–
400	1.31	1.4	1.45	–	–	–
500	1.44	1.52	–	–	–	–
600	1.52	1.56	–	–	–	–

$$S = (1000 - C/\rho_c - W/\rho_w - III/\rho_{cs})\rho_s \qquad (5.2)$$

where ρ_c, ρ_w, ρ_{cs}, ρ_s are values of density for cement, water, crushed stone and sand accordingly; $\rho_{b.cs}$ – bulk density of crush stone.

$$CS = \frac{1000}{1.375 \times \dfrac{0.45}{1.48} + \dfrac{1}{2.68}} = 1264 \text{ kg/m}^3$$

$$S = \left(1000 - \left(\frac{463}{3.1} + 143 + \frac{1264}{2.68}\right)\right) \times 2.65 = 625 \text{ kg/m}^3$$

The calculated concrete has the following composition: cement – 463 kg/m³, water – 143 l/m³, crushed stone – 1264 kg/m³, sand – 625 kg/m³. The consumption of Melflux 2651F superplasticizer is 463 × 0.05 = 23.15 kg/m³.

Example 2: *It is necessary to design the composition of concrete of class C50/60, with slump of concrete mixture Sl = 220 mm, when using ash-content composite cement with a specific surface S_a = 450 m²/kg (see Task 2). Crushed granite stone 5-20 mm with ρ_{cs} = 2700 kg/m³ and $\rho_{b.cs}$ = 1450 kg/m³, medium-sized sand with ρ_s = 2650 kg/m³ and $\rho_{b.s}$ = 1470 kg/m³.*

The design of concrete compositions at the studied composite cement with use of experimental–statistical models can be performed in the following sequence:

1. From the models of concrete strength during hardening under normal conditions (4.5) and after heat moisture treatment (HMT) (4.6), (4.7), we determine the W/C that provides each of the given strength properties.

   To solve these models in relation to W/C, the values of other factors should be set, in particular, the content of SP and the parameters of HMT. In the first approximation, from the condition of minimum cost for concrete, the minimum content of *SP* (0.4%) can be accepted. For cast mixtures, the *SP* content should be increased to 0.7%, for self-compacting concretes to 1.0%. HMT parameters are accepted at the minimum level.

2. From the determined values of *W/C*, the minimum that provides all the properties is accepted. In case of significant (more than 25%) differences between the *W/C* values obtained from different conditions, the content of *SP* and ash in composite cement, as well as the HMT parameters to reduce the 'scissors' in the *W/C*, are changed.

3. Determine the water consumption for the given workability depending on the normalised parameters.

To study the water consumption of concrete mixtures based on composite ash-containing cements three-level, four-factor plan B_4 [42] was used. The conditions for its implementation are given in Table 5.4.

Table 5.4: Conditions for planning experiments

Factors of influence		Levels of variation			Interval
Natural type	Coded type	−1	0	+1	
Content of superplasticizer (*SP*), %	X_1	0.4	0.7	1.0	0.3
The specific surface area of the cement, S_a, m²/kg	X_2	350	450	550	100
Cement consumption *C*, kg/m³	X_3	300	400	500	2
Slump, *Sl*, mm	X_4	20	130	240	110

After processing and statistical analysis of the experimental data, a mathematical model of the water consumption of the concrete mixture is obtained when using ash-containing composite cements:

$$W = 142 - 24.92X_1 + 4.14X_2 + 3.92X_3 + 5.49X_4 -$$
$$0.63X_1X_2 + 0.63X_1X_3 + 1.25X_2X_3 - 1.25X_2X_4 + \qquad (5.3)$$
$$0.38X_3X_4 - 0.68X_1^2 + 1.82X_2^2 + 0.82X_3^2 - 1.18X_4^2$$

Analysing the mathematical model (5.3), we pay attention to the presence of a significant interaction between the effects of the specific surface area of cement and the consumption, as well as the slump of the concrete mixture. In particular, with an increase in S_a, the water consumption of the concrete mixture decreases with an increase in the slump, as well as with a decrease in cement consumption. A significant nonlinear effect can be traced when analysing the influence of the specific surface on the workability of mixtures. The ranking of the quantitative effects of the influence of the studied factors on water consumption allows their placement in descending order of influence $X_1 > X_2 > X_3 > X_4$.

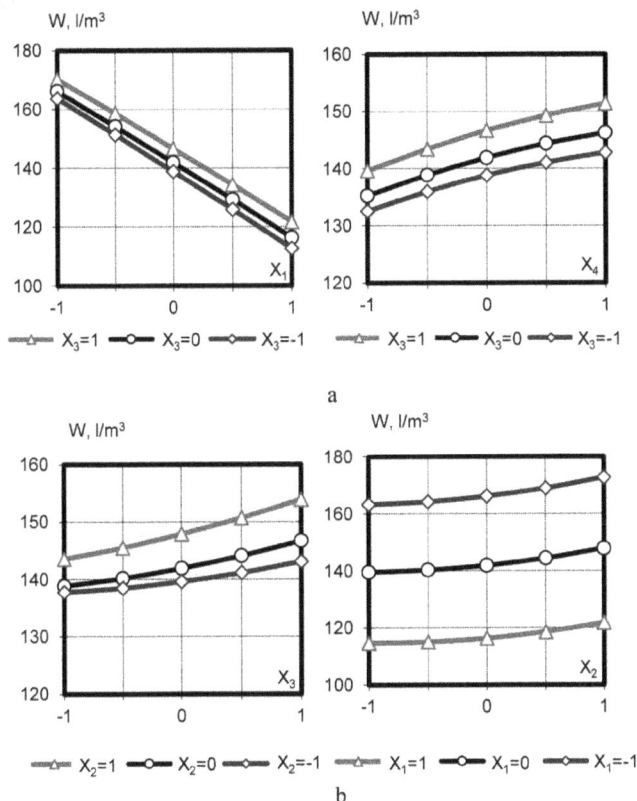

Fig. 5.2: Influence of composition factors on the water consumption of concrete mixtures based on ash-containing composite cements: a – influence of cement consumption (X_3), SP content (X_1) and mixture slump (X_4); b – influence of the specific surface area (X_4), cement consumption (X_3) and SP content (X_1)

Graphic dependencies of the water consumption of the concrete mixture on the investigated technological factors are shown in Fig. 5.2.

Analysing these dependencies, we come to the conclusion that increase in the consumption of superplasticizer from 0.4% to 1% (of the mass of *CC*) causes an almost linear decrease in the water consumption of the concrete mixture from 170 to 180 l/m^3 to 110-120 l/m^3. The influence of other factors is less significant. When the workability of the concrete mixture increases from stiff to cast consistency, water consumption increases by 10-15 l/m^3. The increased dispersion of the binder (450-550 m^2/kg) causes an increase in water consumption by 7-10 l/m^3 compared to the specific surface of 350-450 m^2/kg. The content of *CC* is the least significant factor with regard to the influence on water consumption in the studied range – when the cement content increases from 300 kg/m^3 to 500 kg/m^3, the water consumption increases by 3-6 l/m³, and with increased dispersion up to 10 l/m³.

The analysis of the obtained graphs shows that the workability of concrete mixtures, which corresponds to $Sl > 220$ mm, is achieved with an increased consumption of superplasticizer $SP = 0.7$-1%, S_a – 350-450 m²/kg. Cement consumption, in turn, does not have a significant effect on the concrete slump.

Determining water demand and its regulation from changes in the main technological factors can be carried out with the help of a nomogram (Fig. 5.3), which is built on the basis of the mathematical model (5.3). Nomograms can be used for preliminary selection of composition parameters.

As is known, the water consumption of concrete mixes is correlated with the normal consistency (*NC*) of cement. Corresponding experimental data on the dependence of water consumption of concrete mixtures based on composite cements *NC* with the addition of *SP* at different values of slump are given in Table 5.5.

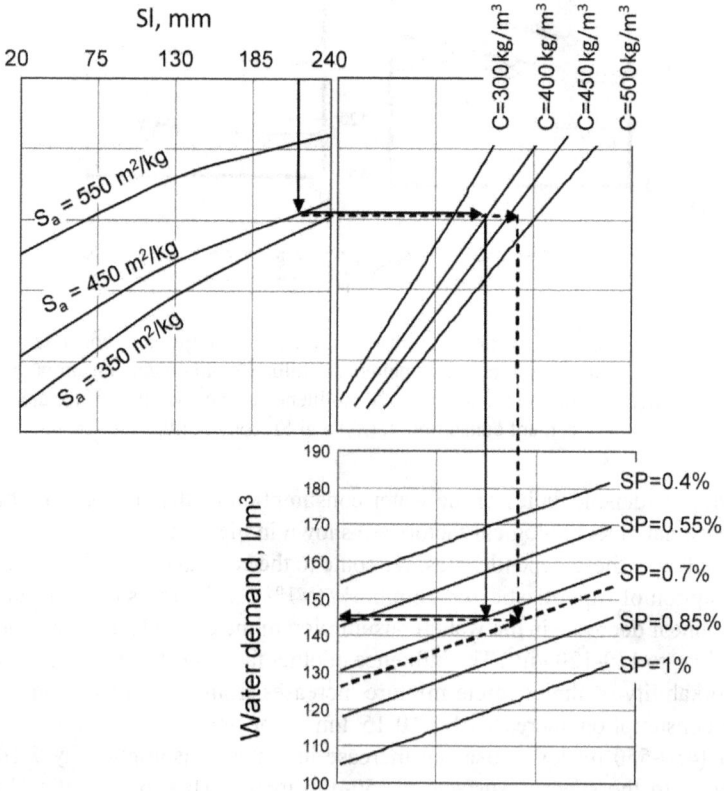

Fig. 5.3: Nomogram for determining water consumption of concrete mixtures, based on ash-containing composite cement (*CC*)

Table 5.5: Dependencies of water consumption of concrete mixtures on *NC* cement and maximum size of aggregates

NC, %	W, kg/m³ at its consistency			
	Sl = 10-40 mm		Sl > 220 mm	
	D_{max} = 20 mm	D_{max} = 5 mm	D_{max} = 20 mm	D_{max} = 5 mm
16	105	135	145	230
20	125	165	175	250
24	150	220	210	275

An amendment to increase water consumption can also be accepted according to Table 5.6.

Table 5.6: Correction for the water consumption of the concrete mixture with cement consumption greater than 400 kg/m³ (*SP* = 0.4%)

Specific surface area of the CC, m²/kg	Consumption of composite cement, kg/m³									
	420	440	460	480	500	520	540	560	580	600
	An increase in the water consumption of the concrete mixture overestimated, kg/m³									
550	4	8	12	15	20	25	30	35	40	45
450	3	6	9	12	15	19	23	27	31	35
350	2	4	6	8	10	12	15	18	21	25

4. Determine the consumption of cement, kg/m³:

$$C = W/(W/C) \qquad (5.4)$$

5. Determine the consumption of aggregates by the method of absolute volumes.

Solution: The required average strength of C50/60 class concrete, determined on cube samples with a coefficient of variation of 13.5%, is 77 MPa. Only the compressive strength is normalised, so we determine the required *W/C* = 0.315 according to the nomogram (Fig. 4.7).

With the initial workability parameter *Sl* = 220 mm and cement dispersity S_a = 450 m²/kg, we set in the first approximation with the approximate consumption of cement *C* = 400 kg/m³ and *SP* admixture = 0.7%. Using the nomogram (Fig. 5.3), determine the water consumption, *W* = 145 l/m³.

Determine the consumption of cement, kg/m³

$$C = 145/(0.315) = 460 \text{ kg}.$$

According to Table 5.3 adjust the water consumption, at *C* = 460 and increase the amount of water by 9 litres. Thus, *W* = 154 litres.

Determine the consumption of cement, kg/m^3:

$$C = 154/(0.315) = 489 \text{ kg.}$$

Determine the consumption of aggregates by the method of absolute volumes:

- Crushed stone consumption, kg/m^3

$$CS = \cfrac{1000}{K_a \cfrac{V_{cs}}{\rho_{b.cs}} + \cfrac{1}{\rho_{cs}}} = \cfrac{1000}{1.37\cfrac{0.46}{1.45} + \cfrac{1}{2.7}} = 1238 \text{ kg/m}^3$$

where K_a is the grain moving apart coefficient (Table 5.3); V_{cs} – voidness of coarse aggregate.

$$V_{cs} = 1 - \frac{\rho_{b.cs}}{\rho_{cs}} = 1 - \frac{1.45}{2.7} = 0.46 \text{ kg/l}$$

- Sand consumption

$$S = \left(1000 - \left(\frac{C}{\rho_c} + W + \frac{CS}{\rho_{cs}} \right) \right) \cdot \rho_s = 609 \text{ kg/m}^3$$

Calculated nominal composition of concrete, kg/m^3: $C = 489$ kg/m^3; at $W/C = 0.315$; $W = 154$ kg/m^3; $SP = 0.007 \times 489 = 3.42$ kg/m^3; $S = 609$ kg/m^3; $CS = 1238$ kg/ m^3.

Example 3: *Calculate the composition of fine-grained high-strength concrete when adding superplasticizer 'Melflux' and metakaolin to the concrete mixture.*

Granite siftings were used as aggregates. To obtain a concrete mixture, Portland cement CEM I is used.

After the factorial experiment, mathematical models of concrete strength and water-cement ratio were obtained, taking into account the consumption of superplasticizer (*SP*), the content of particles less than 0.16 mm in the siftings ($m_{0.16}$) and the content of metakaolin.

After additional calculations, the concrete compressive strength model was reduced to the formula:

$$f_{cm} = AR_c(C/W - b) \tag{5.5}$$

where R_c – cement strength; C/W – cement-water ratio, A, b – coefficients.

Based on the water-cement ratio model, a nomogram was constructed (Fig. 5.4).

The calculation method is as follows:

1. Given the standard compressive strength of cement and the required strength value of concrete at the age of 28 days, choosing the appropriate coefficients from Table 5.7, we find the required C/W:

$$C/W = \frac{f_{cm}}{A \cdot R_c} + b$$

Table 5.7: Coefficients of the equation for determining strength

Content of sifting particles with size <0.16 mm ($m_{0.16}$), %	A	b
0	0.27	-0.79
12	0.38	0.07
24	0.51	0.11

2. Let's go to the W/C:

$$W/C = 1/(C/W) \qquad (5.6)$$

3. According to the nomogram (Fig. 5.4), taking into account the slump of the mixture, the number of particles <0.16 mm in the sifting, find the consumption of cement and superplasticizer.
4. According to formula (5.5), knowing the cement consumption and the water-cement ratio, we find the water consumption:

$$W = C \cdot (W/C). \qquad (5.7)$$

5. According to formulas (5.8-5.10), knowing the consumption and density of cement and the consumption of water, find the volume of cement paste ($V_{c.p}$), the consumption of aggregate by volume (V_a) and mass (m_a):

$$V_{c.p} = \frac{C}{\rho_c} + W \qquad (5.8)$$

$$V_a = 1000 - V_{c.p} \qquad (5.9)$$

$$m_a = V_a \rho_a \qquad (5.10)$$

For example, it is necessary to calculate the composition of fine-grained concrete with a 28-day compressive strength of 60 MPa. The slump of the concrete mixture is 100 mm.

Melflux superplasticizer is used as a plasticizing admixture in the amount of 0.6% of the cement consumption. The standard compressive strength of cement is 50 MPa. The content in the sifting of dusty particles <0.16 mm is 12%, the density of the sifting is 2.7 kg/l.

1. Based on the given strength of cement and the required value of strength of concrete at the age of 28 days (formula 5.5), find C/W:

$$f_{cm} = 0.38 R_c (C/W - 0.07);$$

$$C/W = f_{cm}/0.38 R_c) + 0.07 = (60/0.38 \times 50) + 0.07 = 3.23$$

Cement consumption, kg/m³

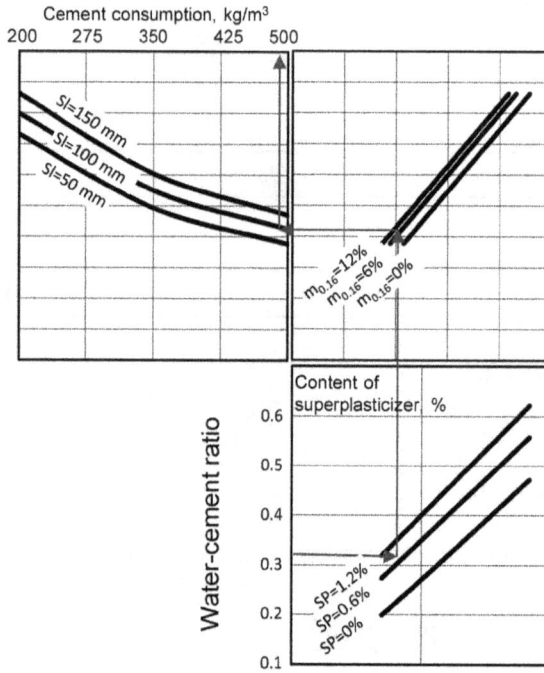

Fig. 5.4: Nomogram for determining cement consumption

2. Find the W/C:

$$W/C = 1/3.23 = 0.31$$

3. According to the nomogram (Fig. 5.4), find the consumption of cement, which is 495 kg/m³.
4. According to the formula (5.7), find the water consumption:

$$W = 495 \cdot 0.31 = 153 \text{ l/m}^3$$

5. Using formulas 5.8-5.9, find the volume of cement paste, volume and mass of aggregate:

$$V_{c.p} = \frac{C}{\rho_c} + W = \frac{495}{3.1} + 153 = 313 \text{ l}$$

$$V_a = 1000 - 313 = 687 \text{ l}$$

$$m_a = 687 \cdot 27 = 1854 \text{ kg/m}^3$$

Design composition of concrete:
- cement – 495 kg/m³
- screening – 1854 kg/m³

- water – 153 l/m^3
- superplasticizer Melflux – 2.9 kg/m^3

Example 4: *Calculate the composition of concrete on activated low-clinker slag Portland cement.*

Cement composition: clinker – 12%, slag – 88%, phosphogypsum – 75%, (SO$_3$ – 4.5%). To ensure sulphate-fluoride-alkaline activation, quicklime in the amount of 3% and sodium silicon fluoride (Na$_2$SiF$_6$) – 2% by mass were additionally added to the composition of the binder.

The standard compressive strength of cement with a specific surface of 453 m^2/kg is 46 MPa. The plasticizing admixture is the polycarboxylate superplasticizer – Sika VC 225. As aggregates, granite crushed stone of fraction 5-20 mm and quartz sand with M_s = 1.9 are used.

The experiments were carried out in accordance with a three-level plan with varying cement consumption X_1 (300-400 kg/m^3) and the content of superplasticizer X_2 (0-3%). The slump of the concrete mixture at all points of the plan was 50-90 mm. Based on the obtained experimental data, models of water consumption (W), compressive strength at the age of seven days (f_{cm}^7) and 28 days (f_{cm}^{28}) at normal hardening and after steaming (f_{cm}^{HMT}, f_{cm}^{HMT28}) were obtained (Table 5.8).

Mathematical models of the compressive strength of concrete at the age of 28 days, in which the water-cement ratio (X_1') and the consumption of plasticizing admixture (X_2') were chosen as variable factors (Table 5.8), allow to calculate the compositions with a given strength and slump of concrete mixture. At the same time, the calculation method is as follows:

1. Choose the type and amount of superplasticizer to ensure the necessary slump of the concrete mixture.
2. According to the specially constructed experimental graphs shown in Fig. 5.5, with the given value of the concrete mixture workability and the determined type and amount of plasticizer, determine the water consumption.
3. To determine the water-cement ratio of the concrete mixture, use mathematical models of the compressive strength of concrete, in which the water-cement ratio (X_1') and the consumption of the plasticizing admixture (X_2') were chosen as variable factors, having previously converted the content of the plasticizing admixture into coded form according to the formula:

$$X_2' = \frac{(SP - 0.3)}{0.3} \tag{5.11}$$

4. Convert the obtained value of the water-cement ratio into its natural form, taking into account that:

$$X_1' = \frac{(W/C - 0.45)}{0.2} \tag{5.12}$$

5. Knowing the water consumption and the water-cement ratio, find the cement consumption:

$$C = \frac{W}{W/C} \qquad (5.13)$$

6. Find the consumptions of aggregates according to the well-known equations [5]:

$$CS = \frac{1000}{K_a \dfrac{V_{cs}}{\rho_{b.cs}} + \dfrac{1}{\rho_{cs}}} \qquad (5.14)$$

$$S = \left(1000 - \left(\frac{C}{\rho_c} + W + \frac{CS}{\rho_{cs}}\right)\right)\rho_s \qquad (5.15)$$

Table 5.8: Experimental-statistical models of water consumption and strength of concrete at the activated cement

Experimental-statistical Models	
Water consumption, l/m³	
$W = 145.3 + 1.667X_1 - 32.507X_2 + 5.894X_1^2 + 11.894X_2^2 - 7.0X_1X_2$	(5.16)
Compressive strength at the age of 7 days, MPa	
$f_{cm}^7 = 17.656 + 7.552X_1 + 6.651X_2 + 6.969X_1^2 + 1.369X_2^2 + 1.2X_1X_2$	(5.17)
Compressive strength at the age of 28 days, MPa	
$f_{cm}^{28} = 42.659 + 12.036X_1 + 12.736X_2 + 1.275X_1^2 - 4.625X_2^2 + 1.2X_1X_2;$	(5.18)
$f_{cm}^{28} = 31.953 - 21.844X_1' - 0.016(X_2')^2 - 0.016(X_2')^2$	(5.19)
Compressive strength after HMT, MPa	
$f_{cm}^{HMT} = 36.076 + 7.451X_1 + 7.968X_2 - 4.123X_1^2 - 2.473X_2^2 - 2.45X_1X_2;$	(5.20)
$f_{cm}^{HMT} = 26.9 - 12.5X_1' + 0.1(X_1')^2 + 0.01(X_2')^2$	(5.21)
Compressive strength after HMT and 28 days of normal hardening, MPa	
$f_{cm}^{HMT\,28} = 54.041 + 14.236X_1 + 15.636X_2 - 2.576X_1^2 - 5.736X_2^2 + 2.55X_1X_2;$	(5.22)
$f_{cm}^{HMT\,28} = 38.5 - 31.9X_1' + 0.1X_2' + 8.3(X_1')^2 + 1.89(X_2')^2 - 0.1X_1'X_2'$	(5.23)

* Heat-moisture treatment of concrete (HMT) was carried out by steaming at a temperature of 80°C. The rate of temperature rise and cooling was 30°C per hour. The duration of isothermal exposure is six hours.

where K_a is the grains moving apart coefficient, $\rho_{b.cs}$ – bulk density of crushed stone, ρ_{cs} – true density of crushed stone, ρ_s – true density of sand, V_{cs} – intergranular voidage of crushed stone.

For example, it is necessary to calculate the composition of concrete made on low-clinker slag Portland cement, characterised by a compressive strength of 35 MPa after heat-moisture treatment and 70 MPa after 28 days of further normal hardening.

The slump of the concrete mixture is equal to 150 mm. Sika VC 225 superplasticizer is used as a plasticizing admixture in the amount of 0.6% of the cement mass.

1. According to the graph shown in Fig. 5, with the specified slump of the concrete mixture (150 mm), the specified type (Sika VC225) and the amount of superplasticizer (Sika VC225 0.6% of the cement mass), determine the water consumption (135 l/m^3).

2. Convert the content of the plasticizing admixture into coded form:

$$X_2' = \frac{(0.6-0.3)}{0.3} = 1$$

3. From Eq. 5.21 and 5.23 (Table 5.8), determine the W/C (X_1'), which will ensure the necessary concrete strength:

- after heat-moisture treatment ($f_{cm}^{HMT} \geq 35$ MPa):

$$35 = 26.9 - 12.5X_1' + 0.1(X_1')^2 + 0.01^2 \cdot 0.1(X_1')^2 - 12.55\,X_1' - 8.1 = 0$$

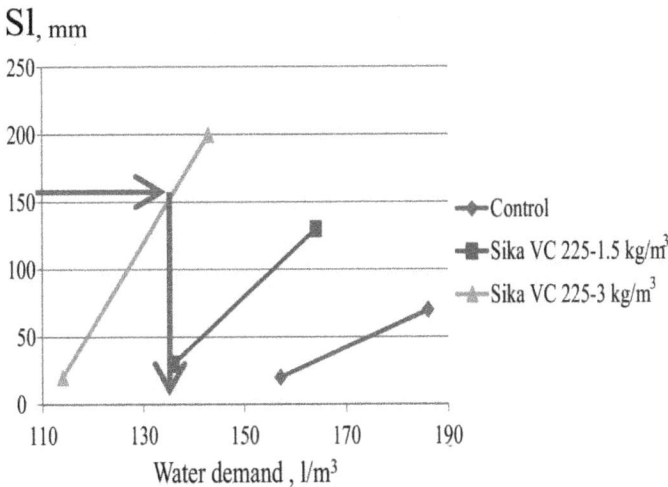

Fig. 5.5: Graphs of the dependence of the slump of the concrete mixture produced on the low-clinker slag Portland cement depending on the water consumption per m^3

After solving the quadratic equation, we get $X_1' = -0.64$.

- after heat-moisture treatment and 28 days of normal hardening ($f_{cm}^{HMT28} \geq 70$ МПа):

$$70 = 38.5 - 31.9\, X_1' + 0.1 \times 1 + 1.89 (X_1')^2 - $$
$$-8.3 (X_1')^2 + 31.8\, X_1' + 33.29 = 0$$

After solving the quadratic equation, we get: $X_1' = 0.86$.

4. Transfer the obtained values of the water-cement ratio into the natural form:

$$W/C = X_1' \times\ 0.2 + 0.45 = -0.64 \times 0.2 + 0.45 = 0.32$$

for $f_{cm}^{HMT28} \geq 70$ MPa $W/C = X_1' \times 0.2 + 0.45 = 0.86 \times 0.2 + 0.45 = 0.28$

To ensure the necessary strength characteristics of concrete, accept the minimum value of the water-cement ratio, which is used for further calculations.

5. Determine the required consumption of cement:

$$C = \frac{135}{0.28} = 482 \text{ kg/m}^3$$

6. Find the consumption of aggregates with a coefficient $K_a = 1.46$ (Table 5.3) (bulk density of crushed stone 1.45 g/cm³, density of crushed stone 2.7 g/cm³, density of sand 2.65 g/cm³, calculated the value between the grain voids of crushed stone is 0.42).

$$CS = \frac{1000}{1.42 \dfrac{0.46}{1.45} + \dfrac{1}{2.7}} = 1218 \text{ kg/m}^3;$$

$$S = \left(1000 - \left(\frac{482}{3.1} + 135 + \frac{1218}{2.7} \right) \right) \times 2.65 = 682 \text{ kg/m}^3$$

The design composition of concrete: cement – 482 kg/m³; water – 135 l/m³; crushed stone – 1218 kg/m³; sand – 682 kg/m³. The consumption of Sika VC 225 superplasticizer is 2.91 kg/m³.

Multi-parameter Design of High-strength Concrete Compositions

General Scheme for Solving Problems

In the most general form, the solution to the problem of designing the composition of the concrete mixture consists in solving the system of equations:

$$P_x = f(P_1 P_2 ... P_m, X_1, X_2 ... X_j) \rightarrow \text{opt};$$

$$P_1 = f_1(X_1 X_2 ... X_k) > P_1^0 (a\partial o < P_1^0)$$

$$P_2 = f_2(X_1 X_2 ... X_k) > P_2^0 (a\partial o < P_2^0)$$

....

$$P_m = f_m(X_1 X_2 ... X_k) > P_m^0 (a\partial o < P_m^0) \tag{6.1}$$

where P_x – optimisation criterion (cement consumption, cost, etc.); P_1, P_2...P_m – parameters of indicators of properties of concrete mixture and concrete; P_1^0, P_2^0...P_m^0 – their normalised values; X_1, X_2...X_κ – parameters of the composition of the mixture.

In construction and technological practice, the methods of designing concrete structures with the necessary compressive strength have become the most widespread. This is due, firstly, to the fact that in structural calculations the strength of concrete is its main parameter, and, secondly, to the assumption that other necessary properties of concrete are unequivocally connected with strength. The latter assumption, however, is not general enough. Indeed, many of its properties are clearly related to the compressive strength of concrete: flexural strength, tensile strength, wear resistance, cavitation resistance, etc. However, the dependencies of strength and frost resistance, strength and creep, etc. are not unambiguous and their calculation should be based on the use of a complex of special quantitative relationships.

The design of concrete compositions can be considered as an isolated system (the first type of problems) and as a subsystem of more general systems; for example, the design of concrete and reinforced concrete structures and their production technologies (the second type of problems). In the first case, the task consists only in the optimal recipe provision of the specified parameters, and in the second, the tasks of optimising the set parameters themselves (workability, strength of concrete, etc.) are additionally solved (Fig. 6.1).

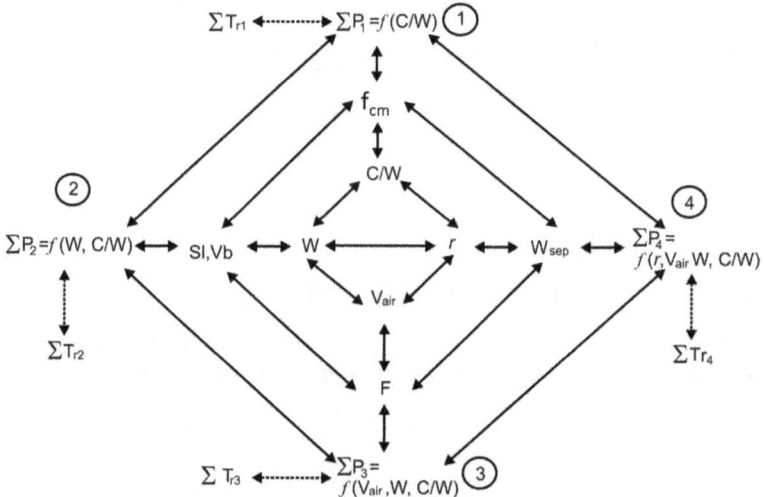

Fig. 6.1: A block scheme for multi-parametric concrete proportioning ($\sum P$ – a group of concrete properties, related to certain parameters of the mixture, $\sum T_r$ – a group of technological factors, affecting concrete properties)

1. Obtaining f_{cm} to achieve $\sum P_1 = f(C/W)$ and required C/W.
2. Obtaining the water consumption W for achieving $\sum P_2 = f(W, C/W)$. Sl or V_b are Slump and Vebe time values, accordingly.
3. Obtaining the entrained air volume V_{air} for achieving $\sum P_3 = f(V_{air}, W, C/W)$.
4. Obtaining the ratio of fine and coarse aggregates (r) for achieving $\sum P_4 = f(r, V_{air}, W, C/W)$.

The design of concrete compositions develops first of all in the direction of taking into account the properties that characterise its durability and compliance with the construction's operating conditions. From the complex of specified properties, which are closely related to the composition of concrete mixtures, it is possible to distinguish deformative properties, resistance of concrete to temperature and humidity effects, corrosion resistance.

Extensive scientific research allows to propose a number of generalised calculation dependencies for taking into account properties that ensure concrete durability in structures when designing concrete compositions.

Accounting for the Deformation Properties of Concrete

The modulus of elasticity (E_c) of concrete is closely correlated with the strength of concrete. When designing structures for predicting the modulus of elasticity of concrete when loaded at the age of τ, are used dependencies of the type:

$$E_c = \frac{E_m f_\tau}{S + f_\tau} \tag{6.2}$$

where f_τ – compressive strength of concrete at a certain duration of hardening (τ); E_m and S – empirical constants whose values are recommended in construction regulations: $E_m = 52000$; $S = 23$.

The European Committee on Concrete and the norms of a number of countries recommend the dependence:

$$E_c = C(f_\tau)^\gamma \tag{6.3}$$

where C and γ – empirical coefficients.

The discrepancies between the values of E_c calculated by formulas 6.2 and 6.3 increase (up to 35%) as the strength of concrete increases.

Various authors have proposed various modifications of formulas and values of the coefficients. A widely used formula for determining the modulus of elasticity of concrete according to US AC1-318-83:

$$E_c = 4540\sqrt{f'_{cm}} \tag{6.4}$$

where f'_{cm} – cylindrical compressive strength of concrete.

As a result of statistical processing of experimental data, a close correlation was established [40] between the elastic modulus of high-strength concretes and their prism strength, described by the dependence

$$E_c = 5370.8\sqrt{f_{pr}}. \tag{6.5}$$

The dependence obtained is practically similar to dependence (6.3) included in the US norms. Concrete strength values up to 200 MPa most closely match the experimental data formula:

$$E_c = k_a k_v k_c k_{sp} k_n k_N \frac{0.05 f_{pr} + 57}{1 + \dfrac{29}{3.8 + f_{pr}}} \tag{6.6}$$

where k_a, k_v, k_c, k_{sp}, k_N – coefficients that take into account the influence of the elastic properties of coarse aggregate, the concentration of coarse aggregate, the influence of the composition of cement and admixtures, as well as the features of national standards in determining the modulus of elasticity.

To calculate the change in the modulus of elasticity of concrete over time [5] empirical dependencies are also proposed, for example,

$$E_\tau = E_0(1 - \alpha \exp(-\beta\tau)) \tag{6.7}$$

where $E_\tau = E_0$ – respectively, the modulus of elasticity of concrete at the age of τ and 28 days, α and β – are empirical coefficients.

The dependence of the modulus of elasticity of concrete should take into account not only on the strength of concrete, but also on a number of other factors characterising its composition, hardening and testing conditions, making it promising to determine it for specific conditions using experimental statistical models.

The elastic properties of concrete can be characterised by both static (E_c) and dynamic modulus of elasticity (E_d), which takes into account the stresses that arise when the sample vibrates. Based on numerous studies, an almost linear relationship has been established between the dynamic and static moduli of elasticity of concrete. For high-strength concretes $E_c/E_d \approx 0.8$.

For concrete, which is subject to increased crack resistance requirements especially in tension, limit strains are normalised. Ultimate compressive strains (ε_{fcm}) are correlated with the prism strength of concrete (f_{pr}):

$$\frac{\varepsilon_{fcm} \cdot 10^{-4}}{f_{pr}} = 0.1 + \frac{11}{f_{pr}} \tag{6.8}$$

An important deformative property of concrete is its ultimate extensibility. Taking into account the complexity of experimental determination, a calculated indicator close in value is used conditional extensibility.

The conditional extensibility $\varepsilon_{c.e}$ is proposed to be found as the ratio of the tensile strength in splitting ($f_{c,tn}$) to the value of the dynamic modulus of elasticity (E_d):

$$\varepsilon_{c.e} = f_{c.tn}/E_d. \tag{6.9}$$

In the process of laboratory control, the value of $\varepsilon_{c.e}$ can be calculated, knowing the tensile strength by splitting and compressive strength:

$$\varepsilon_{c.e} = \frac{f_{c.tn}(1 + 0.07 f_{cm})}{4.10^3 f_{cm}} \tag{6.10}$$

For unknown values of $f_{c,tn}$, using the relation $f_{c\,tn}$ $0.55 f_{cm}^{2/3}$, we can transform dependence (6.5):

$$\varepsilon_{c.e} = \frac{0.1375(1 + 0.07 f_{cm})}{4.10^3 \sqrt[3]{f_{cm}}} \tag{6.11}$$

Dependencies (6.9-6.11) can be used only for tentative estimates of limiting deformations during loading of concrete, including when designing compositions. The significant influence on these parameters of the characteristics of the initial materials and admixtures should be borne in mind. The value of the ultimate tensile strength of concrete also increases markedly with an increase in the content of cement stone and with an increase in the proportion of fine aggregate in the mixture of aggregates.

The ability of concrete to deform over time under prolonged load (creep) has a significant impact on the operation of structures, their durability. For unreinforced concrete at high load levels, creep accelerates the achievement of ultimate strain and failure of the material. At the same time, concrete creep reduces internal stresses caused by shrinkage in homogeneity, which leads to an increase in crack resistance. The creep of concrete is taken into account when predicting the deflection of reinforced concrete columns under the action of constant loads and the possible loss of reinforcement stress in pre-stressed reinforced concrete elements.

A number of empirical formulas have been proposed for calculating the measure of concrete creep, depending on its strength and the main parameters of the composition. Some of the most famous of these formulas are given in Table 6.1.

All formulas for predicting the measure of creep, with the exception of the earliest ones, show its ambiguous relationship with the compressive strength of concrete. Most researchers have proven that the measure of creep is affected by both W/C and the content of cement stone in concrete, which is consistent with the hypotheses about the mechanism of its development in concrete.

Most empirical creep formulas can be reduced to expression (6.12) if the coefficient K in them is represented by some function of W/C and f_{cm}. Expression (6.12) is the simplest, experimentally substantiated and convenient when designing compositions for estimating the measure of creep $C_{m(28)}$, although the possibility of using other calculation formulas (Table 6.1) cannot be ruled out.

When calculating the measure of concrete creep, as shown by many studies, it should be taken into account that at a certain strength, it also depends on the type of cement, the concentration of cement stone, the influence of admixtures and other factors. Temperature conditions and vibration loads during concrete hardening also have a certain influence.

Concrete hardening is accompanied by contraction and hydraulic shrinkage. The greatest impact on the behaviour of concrete in structures is hydraulic shrinkage. Shrinkage deformations cause internal stresses in concrete, which are especially significant when structures dry unevenly and work in cramped conditions. They can cause breaks in the contact zone and the mortar part of concretes and cause cracks, especially in combination with thermal stresses. Shrinkage stresses adversely affect frost resistance, impermeability, fatigue strength, and cause loss of prestress when tensioning the reinforcement.

A.E. Desov, assuming that the samples are deformed during shrinkage evenly over the cross section, proposed the modulus of cement stone cracking, due to shrinkage stresses:

Table 6.1: Basic calculation formulas for predicting a measure creep of normal-weight concrete [5]

No.	Formula	Authors
1	$$C_{m(28)} = \dfrac{K}{f_{cm}} \qquad (6.12)$$ $C_{m(28)}$ – the limiting value of the measure of concrete creep when loaded in 28 days; f_{cm} – cubic compressive strength of concrete at 28 days, MPa; $K = 25 \cdot 10^{-5}$	A. Velmy
2	$$C_{m(28)} = \dfrac{K \cdot W/C(W + 0.33C)}{f_{cm}} \qquad (6.13)$$ W and C – water and cement consumption per 1 m³ of concrete; $K - 1.4 \cdot 10^{-6}$.	A. Velmy
3	$$C_{m(28)} = \dfrac{KV_n}{f_{cm}\sqrt[3]{f_m/f_{cm}}} \qquad (6.14)$$ V_n – pore volume in the cement stone, subject to 15% chemically bound water f_m – ultimate value of cubic strength	A. Hummel
4	$$C_{m(28)} = \dfrac{K \cdot W/C(W + 0\,33C)}{\sqrt{f_{cm}}} \qquad (6.15)$$	European Concrete Committee (ECC)
5	$$C_{m(28)} = K\dfrac{1 + W/C}{1 + W/C + m}(W/C)^2 \qquad (6.16)$$ $K = 11 \cdot 10^{-6}$; m – mass ratio between aggregate and cement.	I.I. Ulitsky
6	$$C_{m(28)} = \dfrac{KW}{f_{cm}} \qquad (6.17)$$ $K = 16 \cdot 10^{-6}$.	E.N. Shcherbakov

$$M_{cr} = f_{ctm}/\varepsilon_{sh} \qquad (6.18)$$

where f_{ctm} is tensile strength of specimens; ε_{sh} – the magnitude of shrinkage deformations by the time cracks appear.

He established that cracking of cement stone is typical at M_{cr} less than 3.5 MPa, shrinkage cracks form in cement stone at absolute shrinkage values from 40 to 150 μm/m.

The well-known empirical formulas proposed for predicting the concrete shrinkage in atmospheric conditions with constant cross-sectional dimensions of the elements differ in the features of taking into account the consumption of water and cement in concrete, as well as their ratio (Table 6.2). The decisive factor determining the shrinkage of concrete is water consumption. At a constant water consumption in the mixture, the value ε_{sh} depends little on the consumption of

Table 6.2: Basic calculation formulas for concrete shrinkage prediction [5]

No.	Formula	Authors
1	$\varepsilon_{sh} \cdot 10^6 = 0.24W^{3/2}\dfrac{(1+C/W)^{3/2}}{6+(C/W)^2}$ (6.19)	E.N. Shcherbakov
	B – water consumption in l/m³; C/W – cement-water ratio.	
2	$\varepsilon_{sh} \cdot 10^6 = 0.125W\sqrt{W}$ (6.20)	E.N. Shcherbakov
3	$\varepsilon_{sh} \cdot 10^6 = \dfrac{5W/C}{1+m}(667+C)$ (6.21)	A. Velmy
	m – mass aggregate and cement ratio.	
4	$\varepsilon_{sh} \cdot 10^6 = 5500\dfrac{1+W/C}{1+W/C+m}(W/C)^2$ (6.22)	European Concrete Committee (ECC)
5	$\varepsilon_{sh} \cdot 10^6 = 300\left[0.7+0.15\left(\dfrac{C-225}{25}+\dfrac{W/C-0.4}{0.1}\right)\right]$ (6.23)	C.V. Aleksandrovskiy

cement and C/W. For engineering calculations, the most simple and convenient formula is (6.20).

In Fig. 6.2 shrinkage deformation curves are given according to our experimental data. The obtained experimental values of shrinkage strains for concretes of various compositions satisfactorily coincide with the calculated values obtained through the formula (6.20).

According to the International Federation for Pre-stressed Concrete (FPC) and the European Committee for Concrete (ECC), after seven days of hardening, shrinkage is 0.2; 28 days – 0.4; 180 days – 0.7; 365 days – 0.8 ε_{sh}.

For reinforced concrete, shrinkage (ε_{sh}) additionally depends on the percentage of structural reinforcement R:

$$\varepsilon'_{sh} = \varepsilon_{sh}(1-10R)_.$$

The durability of concrete is ensured if its composition and structure correspond to the operating conditions of structures and structures. The resistance of concrete is characterized by its ability to maintain the specified quality indicators and performance under the influence of aggressive environmental factors, of which the most characteristic are the effects of temperature and the aquatic environment.

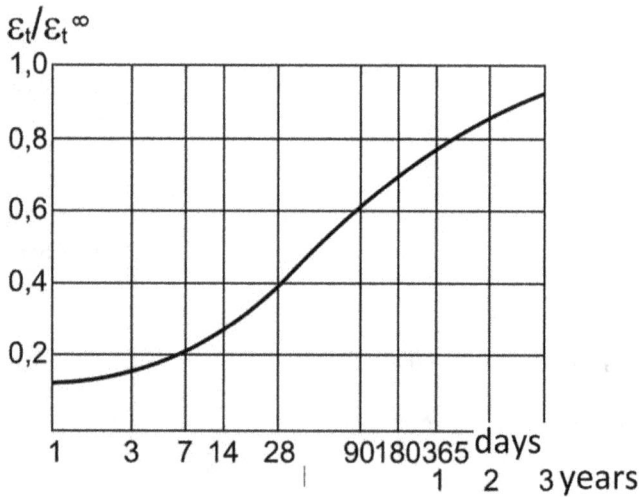

Fig. 6.2: The dependence of shrinkage deformations of concrete in time

Accounting for Frost Resistance and Water Impermeability

Frost resistance of concrete is its ability to maintain strength and performance under the action of alternate freezing and thawing in a state saturated with water. The destruction of concrete in a water-saturated state under the cyclic action of positive and negative temperatures, as well as variable negative temperatures, is due to a complex of physical corrosion processes that cause deformations and mechanical damage to products and structures.

To date, there is no unified theory explaining the mechanism of frost destruction of concrete, although it is obvious that, ultimately, the decrease in the strength of wet concrete during alternate freezing and thawing is mainly due to the formation of ice in the pores of concrete.

The frost resistance of concrete is primarily due to the structure of its pore space. For concrete of normal hardening, justified the dependence:

$$F = K(P_o - P_k)^n \tag{6.24}$$

where F is number of freeze cycles; K, n, P_o are parameters depending on the quality of materials, concrete composition and production parameters; P_k is capillary porosity, %.

Statistical processing made it possible to concretise the given dependence for normal-weight concrete by the empirical equation

$$F = K(14 - P_k)^{2.7} \tag{6.25}$$

The structural criterion for the frost resistance of concrete was proposed by T. Whiteside and X. Sweet. This criterion is known as the 'degree of saturation' (*SD*) and is equal to

$$SD = \frac{V_{f.w}}{V_{f.w} + V_{air}} \qquad (6.26)$$

where $V_{f.w}$ and V_{air} are volumes of freezing water and air per unit volume of concrete.

It was found that at $SD < 0.88$ concrete has high frost resistance, and at $SD < 0.91$ it quickly collapses. Frost resistance F is related to the degree of saturation by an inverse relationship

$$F \approx \frac{1}{SD} = 1 + \frac{V_{air}}{V_{f.w}} \qquad (6.27)$$

The SD parameter makes it possible to qualitatively assess the frost resistance of concrete under various operating conditions. The calculation of the SD value at the design stage of the compositions became possible after the development of theoretical ideas about the porosity of the cement stone and the justification of the corresponding formulas.

G. Falerlund to determine the amount of freezing water (W_f) proposed the following formula

$$W_f = \frac{W/C - 0.25\alpha(0.73 + K_t)}{W/C + 0.32} \qquad (6.28)$$

where K_t is coefficient taking into account the freezing temperature (for $t = -20°C$, $K_t = 0.96$); α is hydration degree.

A criterion of frost resistance (K_F) is proposed, based on the hypothesis that the conditionally closed porosity ($P_{c.c}$) of concrete to prevent its destruction during freezing and thawing should be no less than the possible increase in the volume of water filling the pore space of concrete

$$K_F = \frac{P_{c.c}}{0.09 P_i} \geq 1 \qquad (6.29)$$

where P_i is integral or open porosity, equal to the volumetric water saturation of concrete.

The obvious need to include the volume of entrained air in the criterion of frost resistance led to the appearance of a number of relevant design parameters. These parameters include a modified expression of the so-called 'compensating factor'

$$F_c = \frac{V_{air} + V_c}{V_{fw}} \qquad (6.30)$$

where V_{air} is reserve pore volume formed by entrained emulsified air, %; V_c is volume of contraction pores in concrete, %; V_{fw} is volume of water in concrete that freezes at $-20°C$.

For a qualitative assessment of frost resistance, obviously, there should be a fair condition

$$\frac{V_{air} + V_c}{0.09 V_f} \geq 1 \quad \text{or} \quad \frac{F_c}{0.09} \geq 1 \tag{6.31}$$

Opening the pore volume in the Eq. (6.30) with the help of dependencies connecting them with the degree of hydration α and cement consumption (C), we obtain the expression F_c:

$$F_c = \frac{10 V_{air} + 0.06 \alpha C}{W - 0.5 \alpha C + 1000 (1 - K_c)} \tag{6.32}$$

where K_c is compaction factor.

To calculate the degree of hydration of cement, the relationship established by various authors with the compressive strength of the cement stone can be used. For example, T. Powers presented this dependence in the form of a formula

$$f_{c.s} = 238 \alpha^3, \tag{6.33}$$

where $f_{c.s}$ – compressive strength of cement stone, MPa.

Formulas have also been proposed that make it possible to approximately find α, taking into account the standard strength of cement (R_c)

$$R_c = 110 d^2 \tag{6.34}$$

$$d = \frac{1 + 0.23 \alpha \rho_c}{1 + \rho_c W/C} \tag{6.35}$$

where d is relative density of cement stone; ρ_c – cement density ($\rho_c = 3.1 \dots 3.2$ g/cm^3).

We have found that the dependence of the frost resistance of concrete on the F_c criterion is described by an exponential function of the form

$$F = K(10^{F_c} - 1) \tag{6.36}$$

where K is coefficient depending on the type of cement (for ordinary medium aluminate cement $K = 170$).

When using Portland cement with a content of blast-furnace slag up to 20%, medium-grained quartz sand and crushed granite for a wide range of compositions, the frost resistance of concrete is approximated by the formula

$$F = A_1 f_{cm}^{A_2} \exp^{A_3 V_{air}} \tag{6.37}$$

where $A_3 \approx 0.35$, A_1 and A_2 are coefficients, the values of which change with the change in water content and, accordingly, the workability of the concrete mixture.

The required volume of entrained air in % from the expression (6.37) can be found by the formula

$$V_{air} = \frac{\ln\left(\dfrac{F}{A_1 f_{cm}^{A_2}}\right)}{0.35} \tag{6.38}$$

A quantitative assessment of the influence of individual factors and their interaction is achieved on the base polynomial models, usually obtained by using mathematical planning of experiments. The use of such models makes it possible to implement fairly simple algorithms for calculating concrete compositions with a given frost resistance in combination with other normalized properties.

One of the first polynomial models of the admissible number of cycles F for normal weight concrete is given below:

$$F = 380 - 68X_1 + 162X_2 + 148X_3 - 27X_4 + 22X_5 +$$
$$63X_6 - 39X_2^2 + 50X_1X_2 + 27X_2X_3 + 24X_3X_6$$
(6.39)

where $X_1 = (W - 180)/30$; $X_2 = (C/W - 2.1)/0.8$; $X_3 = (V_r - 0.06)/0.06$; $X_4 = (NC - 27.2) \times 2.6$; $X_5 = (R_c - 41.2) \times 6.7$; $X_6 = (lg\tau - 1.86) \times 0.41$.

In the given coded values of the factors, W is the water consumption, l/m^3; C/W – cement-water ratio; NC – normal consistency of cement, %; R_c is the standard compressive strength of cement, MPa; τ is the duration of hardening, days; V_r is consumption of air-entraining admixture (Vinsol resin).

When using polynomial models, one should take into account their locality, the possibility of using them within the studied factor space.

The water permeability of concrete is evaluated by four indicators – the levels corresponding to the maximum water pressure (0.1 MPa), at which, according to visual assessment, no water filtration occurs on at least four out of six samples ('wet spot method'); filtration coefficient – the amount of water penetrating through a unit of time with a gradient (the ratio of head in meters of water column to the thickness of the structure in metres) equal to 1; depth of water penetration; breathability.

The ratio of the results obtained using the three methods discussed above for determining water permeability is given in Table 6.3.

Table 6.3: The ratio of concrete grades for water permeability, filtration coefficient and depth of penetration pressurized water

Method for determining water permeability	*Impermeability levels*					
	Normal		*Reduced*	*Low*	*Especially low*	
Concrete level of water permeability	P2	P4	P6	P8	P10-P14	P16-P20
Permeability coefficient filtration, cm/s	At $7 \cdot 10^{-9}$ to $2 \cdot 10^{-8}$	At $2 \cdot 10^{-9}$ to $7 \cdot 10^{-9}$	At $6 \cdot 10^{-10}$ to $2 \cdot 10^{-9}$	At $1 \cdot 10^{-10}$ to $6 \cdot 10^{-10}$	At $5 \cdot 10^{-11}$ to $1 \cdot 10^{-10}$	Less $5 \cdot 10^{-11}$
Depth of penetration of water under pressure, mm	More 150	More 150	At 60 to 150	At 35 to 60	At 20 to 35	Less 20

An accelerated method for determining the water permeability of concrete by its air permeability is also used.

Concrete water impermeability level corresponds in a certain range of concrete air penetration (Table 6.4).

Table 6.4: Relationship between air penetration and water impermeability

Air penetration of concrete, s/cm³	Concrete levels for water impermeability, P	Air penetration of concrete, s/cm³	Concrete levels for water impermeability, P
3.1-4.5	P 2	19.7-29	P 12
4.6-6.5	P 4	29.1-42	P 14
6.6-9.4	P 6	42.1-60.9	P 16
9.5-13.7	P 8	61-88.5	P 18
13.8-19.6	P 10	88.6-130.2	P 20

The relationship between the filtration coefficient of concrete and its compressive strength is experimentally substantiated (Fig. 6.3). The presence of this connection made it possible to propose a method of determining the water impermeability of concrete based on its compressive strength. Experimental data are well approximated by a power dependence of the type:

$$K_f = K_w f_{cm}^m \qquad (6.40)$$

where K_w and m – coefficients, the value of which is influenced by the characteristics of the composition of concrete mixtures, the conditions and duration of hardening, etc.

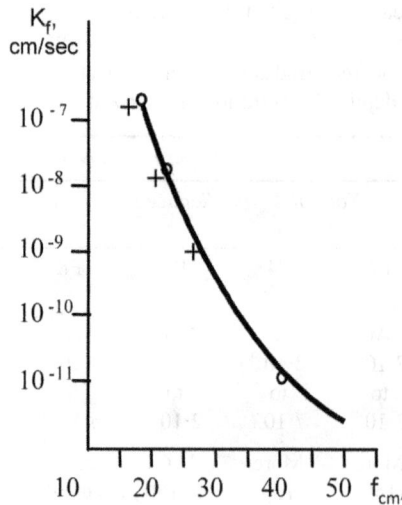

Fig. 6.3: Dependence of concrete filtration coefficient on compressive strength

In view of the great variety of factors affecting the water permeability of concrete, the coefficients K_w and m in formula (6.40) must be specified experimentally when solving the tasks of designing concrete compositions.

V.P. Sizov proposed to predict the impermeability (I) of concrete according to the formula:

$$I = A(R_c/10)\,(C/W - 0.5) \tag{6.41}$$

where R_c is standard cement compressive strength, MPa.

Coefficient A for the reference composition is proposed to be equal to 1, and for other compositions to be specified according to a special nomogram that takes into account the influence of W/C, consumption of water, active and inert admixtures, volume of air pores.

In some cases, in practice, it is necessary to evaluate the change in the permeability coefficient of concrete during the long-term effect of water pressure, that is, take into account the self-compaction of concrete. Self-compaction of concrete in natural conditions has been noted by many authors. Thus, the data on the water permeability of concrete, determined on cores drilled at different times from Italian dams, indicate a decrease in the this coefficient over the period from 90 days to 27 years by 3 orders of magnitude.

The concrete filtration coefficient over time when $\tau > 1$ day ($K_{f(\tau)}$) is described by an exponential function:

$$K_{f(\tau)} = K_{fo}e^{-(\tau/T)^b} \tag{6.42}$$

where K_{fo} is initial coefficient permeability, determined on the first day of water percolation through concrete, m/s; b, T are empirical parameters of the self-compaction function depending on the composition of concrete, hydrocarbonate stiffness of water and pressure gradient.

Features of High-strength Concrete Composition Multi-parameter Design

The use of the generalised equations given above makes it possible to quickly predict the change in the main properties of concrete when its composition changes and to design composition of concrete with the given values of its properties. At the same time, the presence of empirical coefficients in the majority of calculation formulas, which require additional experimental refinement, insufficient consideration of factors important in a particular problem, makes it expedient to use polynomial experimental-statistical models.

In those cases when, in addition to compressive strength, there is a need to normalize a number of its other construction and technical properties, the design tasks of the composition become significantly more complicated. Such problems of multi-parametric design arise when designing compositions of various, and especially, special types of concrete (high strength, hydraulic, road, corrosion-resistant, etc.).

Figure 6.4 shows an example of the relationship between creep and the content of cement stone in concrete at f_{cm} = const. The creep measure of concrete was calculated according to the formula (6.15) proposed by the European Concrete Committee.

From formula (6.15) and Fig. 6.4 it follows that with a constant W/C and therefore strength of concrete, its creep can differ significantly depending on the content of cement stone in concrete. Similarly, it is possible to show the ambiguity of the dependence of the strength of concrete with a group of properties depending mainly on capillary porosity (water absorption, frost resistance, etc.).

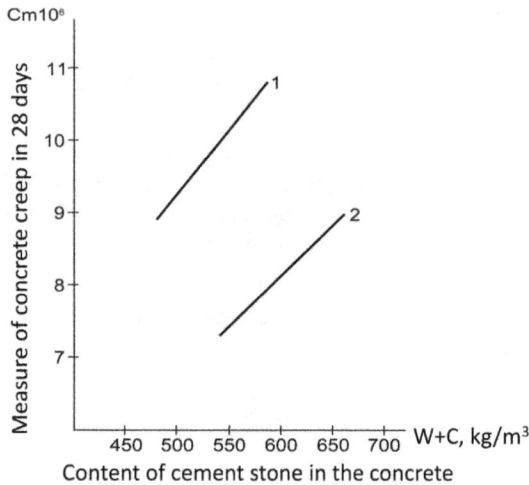

Fig. 6.4: The influence of the content of cement stone in concrete on the degree of creep: $1 - f_{cm} = 20$ MPa; $2 - f_{cm} = 30$ MPa

In order to solve problems of this type, the area of W/C or C/W that provides normalised parameters is established, technological ways of its narrowing are considered, and finally the necessary value of the required W/C is considered. Adjusting the normalised W/C in these tasks is expedient when other composition factors are simultaneously changed, in particular the amount of cement paste, the volume of entrained air, etc.

For example, the average compressive strength of concrete $f_{cm} = 65$ MPa and measure of creep $C_m \times 10^6 = 3.5$ are normalised. The slump of the concrete mixture on granite crushed stone and medium-grained quartz sand is taken as $Sl = 20$ mm. Cement compressive strength $R_c = 50$ MPa. According to the calculation formula of concrete strength $f_{cm} = AR_c (C/W - 0.5)$ with $A = 0.6$, $C/W = 2.63$ ($W/C = 0.38$). When water consumption $W = 175$ l/m^3 and, accordingly, cement consumption $C = 460$ kg/m^3, the required value of the creep is not ensured ($C_m \times 10^6 = 4.5$). In order to achieve the normalised value of $C_m \times 10^6$, it is necessary to increase C/W and, accordingly, reduce the value of W/C. A practical coincidence of W/C from the conditions of strength and creep can be achieved by transition to a stiffer mixture that requires less water content and, accordingly, provides less W/C.

A powerful means of reducing 'scissors' along the W/C in frost-resistant concrete is the entraining of air. Characteristically, while significantly increasing the possible values of W/C to achieve the specified frost resistance, the entrained air at the same time reduces the necessary W/C from the strength condition. At the same time, the overall positive effect of reducing the consumption of cement can be quite significant, especially in concrete with high values of frost resistance at a moderate normalised strength value. From Fig. 6.5, in particular, it follows that $f_{cm} = 20$ MPa and F 200 are provided without the addition of entrained air at $W/C = 0.5$ with the introduction of entrained air – $W/C = 0.62$. At the same time, the value of 'scissors' according to W/C, which is necessary to ensure the specified strength and frost resistance of concrete in the first case is $1.15 – 0.5 = 0.65$, and in the second $0.92 – 0.62 = 0.3$.

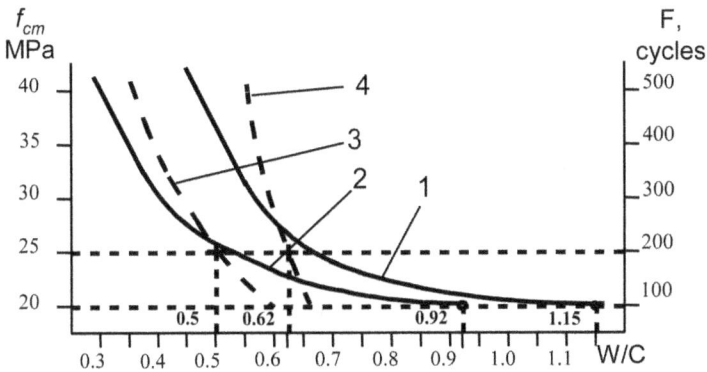

Fig. 6.5: W/C change depending on the given values of strength and frost resistance of concrete: 1. f_{cm} curve without entrained air; 2. f_{cm} with 20 l entrained air; 3. Curve of frost resistance of concrete without entrained air; 4. Frost resistance of concrete with 20 l entrained air

Similarly, 'scissors' in terms of water consumption form, for example, indicators of the workability of the concrete mixture and concrete shrinkage (Fig. 6.6). This requires the inclusion of special calculations related to the determination of such compositions that provide the entire set of normalised properties into the algorithms of multi-parametric design of concrete compositions. Reducing the 'scissors' in terms of C/W, W, r, the content of admixtures determined by the normalised concrete quality parameters and shifting them to the side that provides the best values of the selected criterion (cement consumption, concrete cost, etc.) is the goal of optimisation in multiparameter design compositions of concrete mixtures.

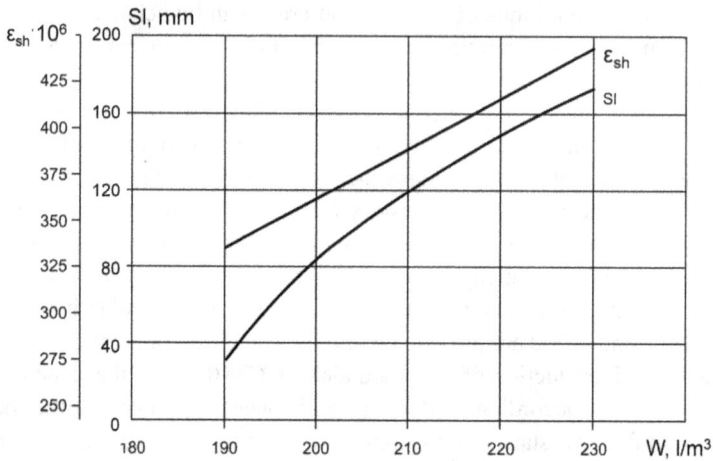

Fig. 6.6: The effect of water content on the workability of the concrete mixture (*Sl*)
and shrinkage (ε_{sh})

Note: Dependence of *Sl* on *W* is accepted for ordinary materials; concrete shrinkage ε_{sh} is
calculated according to the formula $\varepsilon_{sh} \cdot 10^6 = 0.125 \, W\sqrt{W}$

Algorithms for solving the problems of multiparametric design of concrete
compositions are based on the basic laws of concrete science and use of additional
empirical dependencies.

The general scheme of these algorithms is as follows:

1. Taking into account the design requirements for concrete, technological
   conditions and technical and economic analysis, the initial components of
   the concrete mixture and its workability are selected.
2. In those cases when the properties of concrete are normalised, which are
   uniquely related to the strength of concrete under compression (tensile
   strength, flexural strength, modulus of elasticity, conditional extensibility,
   etc.), the value of the strength, which provides the specified properties, is
   determined.
3. Taking into account the compressive strength of cement, quality
   characteristics of aggregates, hardening conditions and other factors, C/W
   is determined, which ensures the specified properties.
4. In order to achieve the necessary level of workability and, if necessary,
   other properties of the concrete mixture and concrete (for example,
   shrinkage) when using these initial materials and admixtures, the water
   consumption (W) is determined. At the same time, in case of going beyond
   the limits of the rule of water demand constancy, the water consumption is
   adjusted taking into account C/W.
5. When normalising the frost resistance of concrete, the required volume of
   entrained air is calculated and the required C/W is specified.

6. With the found values of W and C/W, the possibility of achieving the normalised properties determined by these two technological parameters is checked. In case of non-attainment of normalised parameters, additional correction of W and C/W is carried out using, if necessary, special technological methods (introduction of admixtures, etc.).

7. The consumption of cement is calculated on the basis of the finally found C/W and W and the compliance with the restrictions related to the consumption of cement (heat release, resistance to corrosion, etc.) is checked.

8. The necessary grain composition of fine and coarse aggregate is calculated when several fractions are introduced and then, their consumption. When choosing the ratio of aggregates, along with achieving the best workability and strength, other conditions are also taken into account (increased impermeability, construction thickness, degree of reinforcement, etc.).

9. The possibility of using various technological solutions aimed at saving cement, reducing energy consumption and the cost of concrete mixture is considered.

Example 1: *Calculate the composition of concrete with design compressive strength* f_{cm} = 80 *MPa, shrinkage deformations* $\varepsilon_{sh} \leq 0.3 \cdot 10^{-3}$, *conditional extensibility* $\varepsilon_{c.e} \geq 1.7 \cdot 10^{-4}$. *Slump of concrete mixture Sl* = 80 – 100 *mm.*

Raw Materials: Portland cement with compressive strength R_c = 42.5 MPa, normal consistency NC = 27.5%; quartz sand with a fineness modulus M_f = 1.7, the content of dusty and clay impurities 1.5%, density ρ_s = 2.67 g/cm³; crushed stone fraction 5–70 mm with the content of dusty and clay impurities 1.2%, density $\rho_{c.s}$ = 2.65 g/cm³.

1. Find, using formula 1.14 and Table 5.1, the necessary values of W/C, consumption of water W and cement C to ensure the specified strength of concrete and slump of the concrete mixture. Calculate the consumption of cement

$$W/C = 0.28; \ W = 200 \text{ l/m}^3; \ C = 714 \text{ kg/m}^3$$

2. Using formula 6.20, we will find whether the normalised shrinkage of concrete is ensured at a water content of W = 200 l/m³:

$$\varepsilon_{sh} = 0.125 \cdot 200 \cdot 14.4 \cdot 10^{-6} = 0.36 \cdot 10^{-3}$$

Taking into account that the normalised shrinkage ε_{sh} = 0.3·10⁻³, as well as the consumption of cement is more than 600 kg/m³ (the maximum recommended value for normal weight concrete), we will use a polycarboxylate-type super-plasticizer admixture to reduce water consumption, which allows, at a dosage of 0.7% of cement mass, reduction of water consumption by 35% and correspondingly consumption of cement (for example, Mapei Dynamon SP-3). Then:

$$W = 130 \text{ l/m}^3; \quad \varepsilon_{sh} = 0.125 \cdot 130 \cdot 11.4 \cdot 10^{-6} = 0.19 \cdot 10^{-3}$$

Assuming that the introduction of a superplasticizer does not reduce the design strength at constant W/C, we specify the consumption of cement:

$$C = 130/0.28 = 464 \text{ kg/m}^3$$

3. According to formula 6.11:

$$\varepsilon_{c.e} = \frac{0.1375(1+0.07 \cdot 80)}{10^3 \sqrt[3]{80}} = 2.1 \cdot 10^{-4}$$

Thus, the calculated value of $\varepsilon_{c.e}$ turned out to be higher than the norm.

Example 2: *Calculate the composition of high-strength SCC (self-compacting concrete) (Slump – 250-270 mm) using ground blast furnace slag with a specific surface area of 270 m²/kg and polycarboxylate superplasticizer. The required indicators are concrete class C50/60 (cubic strength at 28 days at least 77.1 MPa), cubic strength and modulus of elasticity at one day must meet the normative indicators for class C12/15 – at least 19.3 MPa and 27000 MPa, respectively.*

Characteristics of Raw Materials: Densities of cement, slag, sand and crushed stone are 3100 kg/m³, 2900 kg/m³, 2650 kg/m³, 2700 kg/m³. The volume part of crushed stone in the SCC should not exceed 0.35.

In order to design composition of *SCC* with blast furnace slag, taking into account a set of specified properties, it is important to obtain mathematical dependences allowing evaluation of technological and concrete composition factors' impact on the required properties. The method of mathematical experiments planning [40-43] is effective to obtain such dependences.

As variable factors were selected, Portland cement consumption was (C, kg/m³ (X_1)), superplasticizer content (SP, % (X_2)) and consumption of mineral admixtures (blast furnace granulated slag) (slag, kg/m³ (X_3)). The concrete mixture workability at the points of the experiment was maintained at the level of $S5$ (slump – 250-270 mm, which corresponds to the cone spread of 560-620 mm ($F5$ according to [45]). 204 cubic specimens of 100×100×100 mm and 48 prisms of 100×100×400 mm were prepared from the concrete mixture. Compressive strength (f_{cm}, f_{prm}, MPa) and modulus of elasticity (E_{pr} were determined at one and 28 days. Mechanical properties of concrete (strengths of cubes and prisms) under a single short-term load were determined by standard methods [46].

For determining the strength of concrete cubic specimens, the experiment was performed according to a three-level three-factor plan B_3 [42] to obtain mathematical polynomial models of the form:

$$Y = b_0 + b_1 X_1 + b_2 X_2 + b_3 X_3 + b_{11} X_1^2 + b_{22} X_2^2 +$$
$$b_{33} X_3^2 + b_{12} X_1 X_2 + b_{13} X_1 X_3 + b_{23} X_2 X_3 \qquad (6.43)$$

where Y is the output parameter, X_1-X_3 are variables, b_0-b_{23} are the equation coefficients.

Investigation of deformation characteristics was carried out in accordance with the two-level plan of the three-factor plan [42]. The type of mathematical model is:

$$Y = b_0 + b_1X_1 + b_2X_2 + b_3X_3 + b_{12}X_1X_2 + b_{13}X_1X_3 + b_{23}X_2X_3 \qquad (6.44)$$

The planning conditions are given in Table 6.5.

Table 6.5: Experiment planning conditions

No.	Factors		Variation Levels			Variation interval
	Natural	*Coded*	*−1*	*0*	*+1*	
1	Cement consumption (C, kg/m³)	X_1	200	400	600	200
2	Superplasticizer content (SP, %)	X_2	0	0.5	0.1	0.5
3	Consumption of blast furnace granulated slag (slag, kg/m³)	X_3	0	100	200	100

The following materials were used in the research: CEM I 42.5 Portland cement, sand with M_f = 2.0, locally available granite crushed stone fraction 5-20 mm, blast furnace granular slag, polycarboxylate type superplasticizer PCE50 (LLC UA-Chemical, Ukraine). The experiment planning matrix and the experimental results are given in Table 6.6.

By statistical analysis of the experimental results (Table 6.6), mathematical models of the output parameters, as well as correlations between individual initial parameters were obtained. The coefficients of mathematical models are given in Table 6.7.

SCC with organic-mineral modifier, which includes polycarboxylate SP and blast furnace granulated slag, after 1 day of hardening had compressive strength in the range of 3.6 to 51 MPa. As expected, factors X_1 (cement consumption) and X_2 (SP content), which cause a significant decrease in W/C, contribute to a significant increase in strength at one day.

The optimal content of the SP admixture is significantly related to the cement consumption – at a minimum content of C just 0.5% of SP is enough to maximise the strength, at a maximum cement consumption the optimal SP content is 0.8-1.1. With increase in the slag content, the optimal content of the admixture and its effectiveness slightly decrease (Fig. 6.7).

Increase in the blast furnace slag content (X_3) mostly reduces the strength at one day of SCC hardening by 40-70%. As known [47], blast furnace slag reacts rather slowly with cement hydration products (primarily with portlantide), so a greater effect should be expected at longer hardening duration. At low cement consumptions, blast furnace slag plays a positive role – there is an increase in f_{cm1} by 20-25%.

Table 6.6: The results of determining the experiment output parameters

No.	Coded factors			W/C	Compression strength, MPa		Prismatic strength, MPa		Modulus of elasticity, MPa	
	X_1	X_2	X_3		f_{cm1}	f_{cm28}	f_{prm1}	f_{prm28}	E_1	E_{28}
1	1	1	1	0.26	25.2	81.1	22.7	45.0	25529	47563
2	1	1	-1	0.25	53.8	103.9	31.8	85.0	27335	53612
3	1	-1	1	0.56	9.9	40.4	9.2	35.0	12946	27102
4	1	-1	-1	0.44	13.3	44.7	9	32.5	20022	35270
5	-1	1	1	0.63	9	51.8	7.5	50.0	16545	43033
6	-1	1	-1	0.73	4	20	2.25	10.7	9165	9947
7	-1	-1	1	1.08	2.8	14.3	2.25	11.7	14709	16633
8	-1	-1	-1	0.86	2.7	14.4	2.5	13.6	10697	28078
9	1	0	0	0.29	43.3	92.3	-	-	-	-
10	-1	0	0	0.56	10.3	46.7	-	-	-	-
11	0	1	0	0.29	23	90.8	-	-	-	-
12	0	-1	0	0.55	9.3	43.7	-	-	-	-
13	0	0	1	0.36	17.5	53.5	-	-	-	-
14	0	0	-1	0.30	29	76	-	-	-	-
15	0	0	0	0.31	22	71.3	-	-	-	-
16	0	0	0	0.31	22.4	70.9	-	-	-	-
17	0	0	0	0.31	21.8	70.5	-	-	-	-

Table 6.7: Coefficients of experimental-statistical models of output parameters

	b_0	b_1	b_2	b_3	b_{11}	b_{22}	b_{33}	b_{12}	b_{13}	b_{23}
	Mathematical model according to Eq. (6.43)									
W/C	0.30	-0.21	-0.13	0.03	0.13	0.13	0.04	0.01	0	-0.05
f_{cm1}	23.6	11.67	7.71	-3.85	1.98	-8.67	-1.54	6.06	-4.63	-2.53
f_{cm28}	74.3	21.51	19.00	-1.79	-6.81	-9.01	-11.56	7.11	-7.34	1.69
	Mathematical model according to Eq. (6.44)									
f_{prm1}	10.9	7.3	5.2	-0.5	-	-	-	3.9	-0.5	-0.5
f_{prm28}	35.5	13.9	12.2	-0.03	-	-	-	3.5	-9.3	-0.144
E_1	17119	4340	2525	314	-	-	-	2449	-2534	1080
E_{28}	32655	8232	5884	928	-	-	-	3817	-4482	5831

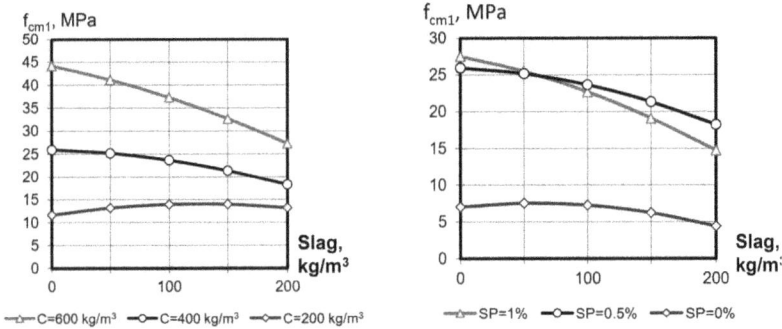

Fig. 6.7: Dependence of compressive strength at one day on the blast furnace slag content

The concrete classes by strength at 28 days that were investigated in the frame of the present research ranged from C 8/10 to C 70/85. In general, the tendencies of the factors affect detected for strength at one day are similar to those at 28 days (Table 6.4, Fig. 6.8). It should be mentioned that the efficiency of factor X_3 (blast furnace slag consumption) becomes higher. The extreme influence of blast furnace slag content becomes more noticeable, making it possible to determine its optimal amount from the position of achieving maximum strength: at a cement consumption of 600 kg/m^3 the optimal slag content from the strength viewpoint is 40-50 kg/m^3 and it is 120-150 kg/m^3 at cement consumption of 200 kg/m^3.

Prismatic strength and modulus of elasticity of SCC with organic-mineral modifier were studied by testing 100×100×400 mm prisms. The influence of the same three factors (cement consumption (X_1) C, kg/m^3, superplasticizer content (X_2) SP, %, blast furnace granular slag consumption (X_3) slag, %) was studied according to the linear plan of the full factorial experiment [48]. The coefficients of the mathematical model are given in Table 6.7.

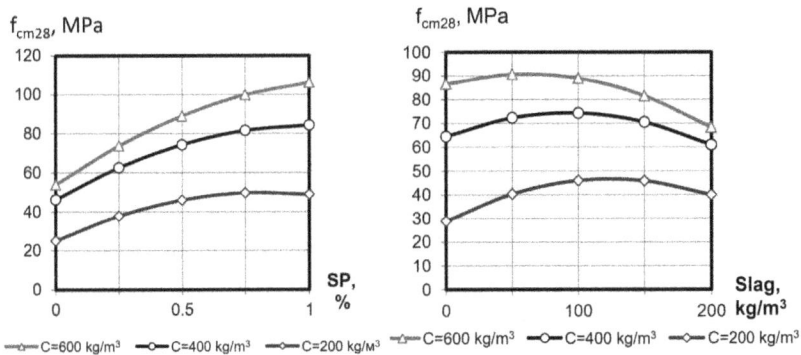

Fig. 6.8: The influence of SP and blast furnace slag contents at different levels of cement consumption on the SCC compression strength at 28 days (the value of the third factor in the graphs is at the average level (Table 6.5)

The prismatic strength after one day of hardening ranged from 2 to 32 MPa, which corresponded to 78% of concrete cubes strength at the same age. The highest influence on the prismatic strength is caused by factors X_1 (C) and X_2 (SP), which first of all cause a decrease in W/C in transition from the lower to the upper level (Fig. 6.9). Factor X_3 (content of blast furnace granulated slag) has little effect on the prismatic strength after one day of hardening.

After 28 days of hardening (Fig. 6.10) interaction of factors X_3 (slag content) with X_1 (cement consumption) becomes more noticeable: for low cement consumption increase in slag content has a positive effect on concrete prismatic strength, while at high cement consumption the same increase in slag consumption causes a decrease in strength. As known [5], concrete deformations play a greater role in concrete prisms destruction (compared to the test of cubes), so the

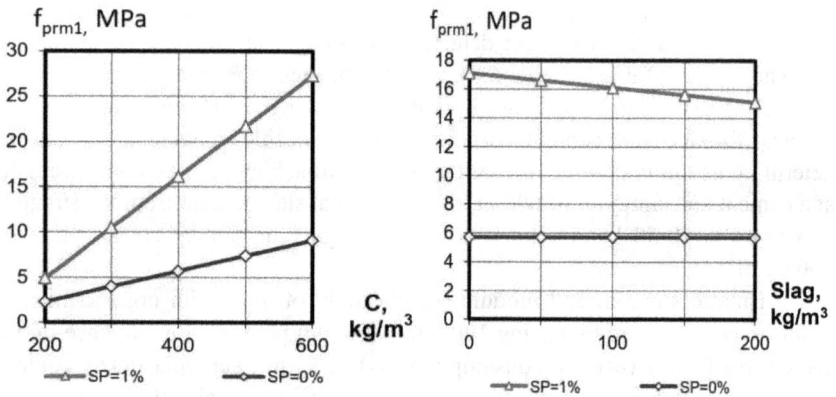

Fig. 6.9: Influence of varied factors on *SCC* prismatic strength at one day (the value of the third factor on the graphs is at the average level (Table 6.5))

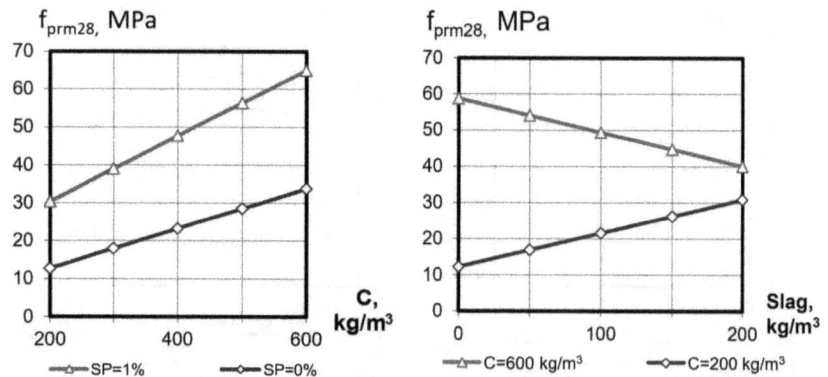

Fig. 6.10: Influence of varied factors on *SCC* prismatic strength at 28 days (the value of the third factor on the graphs is at the average level (Table 6.5))

prismatic strength can be considered more as concrete deformation characteristic. The positive effect of cement slag component at high values of W/C also confirms the structural role of low-basic calcium hydrosilicates, which are formed to a higher extent than in clinker cements.

Influence of other factors (cement consumption and SP content) on the prismatic strength at 28 days is similar to that after 1 day of hardening. The strength varies from 11 to 85 MPa.

The values of concrete modulus of elasticity at the first day were in the range from 10,000 to 27,000 MPa, at the 28th day – from 19,000 to 50,000 MPa. Analysing the mathematical models, it should be noted that all three investigated factors significantly affect the *SCC* modulus of elasticity. The interactions between these factors are also significant.

At transition of factors X_1 (C, kg/m^3) and X_2 (SP, %) from the lower level to the upper one the *SCC* modulus of elasticity increases on average from 10,000-12,000 to 20,000-26,000 MPa at one day and from 20,000-30,000 to 40,000- 50,000 MPa at 28 days. The positive interaction coefficient of these factors (b_{12}, Table 6.7) indicates a significant increase in the modulus of elasticity values with a simultaneous increase in cement consumption and superplasticizer content, both after one day and after 28 days of hardening. Factor X_3 (slag consumption) on average has a less effect on the modulus of elasticity than X_1 and X_2, but its effect is significantly enhanced by the simultaneous influence of other factors. At low cement consumption ($C = 200$ kg/m^3) the increase in slag consumption increases E_{pr} by 1.5-1.7 times both at early hardening stages and at 28 days, however at high cement consumption the same factor causes significant reduction of *SCC* modulus of elasticity (Figs. 6.11 and 6.12).

The action of superplasticizer has a significant influence on the change of *SCC* modulus of elasticity vs. the blast furnace slag content: if in compositions without *SP* positive effect of factor X_3 is manifested only at low cement consumption, then the diluting and water-reducing action of polycarboxylate SP contributes to the positive effect of the slag component and at higher cement consumption it probably levels the significant increase in concrete mixture water demand, causing formation of a denser concrete structure.

A method of creating and using multifactor dependencies obtained, for example, by mathematical experiments planning has a higher resolution to perform tasks of predicting concrete properties at the composition design stage. The obtained experimental-statistical models with coefficients given in the Table 6.7 enable to predict a set of strength and deformation properties at certain values of the investigated factors: consumptions of cement, blast furnace slag and superplasticizer. The use of mathematical optimisation methods [42] and the corresponding software allows finding the values of the above-mentioned factors that provide the necessary properties with minimal resource consumption [49, 50].

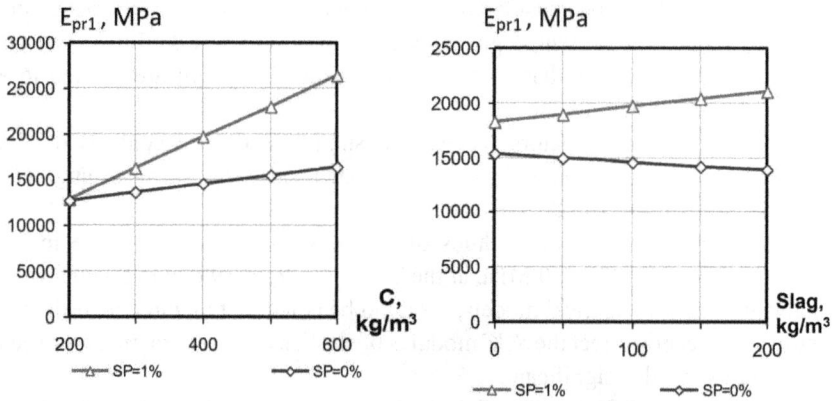

Fig. 6.11: Influence of cement, slag and *SP* contents on *SCC* modulus
of elasticity at one day

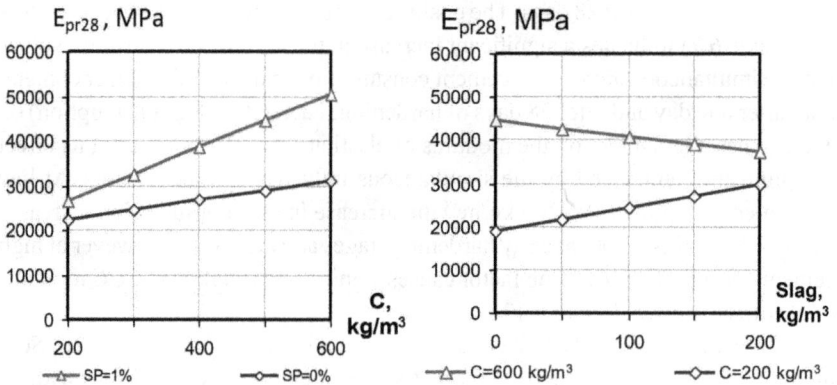

Fig. 6.12: Influence of cement, slag and *SP* contents on *SCC* modulus
of elasticity at 28 days

Solution:

1. According to mathematical models of f_{cm1}, f_{cm28}, E_1 (see Table 6.7) using MS Excel 'Solver' tool find the values of factors (consumptions of cement and slag, and superplasticizer content) that provide required indicators and satisfy the condition that the total cost of these components is minimal (Table 6.8). For calculations purposes the following cost of components was assumed: cement – 88 EUR/t, ground blast furnace slag – 27 EUR/t, superplasticizer – 2.85 EUR/t. These cost values were taken on the basis of market prices in force at Ukraine in 2021.

2. According to the mathematical model of *W/C* (see Table 4) calculate the value of *W/C* at the optimal factor values and concrete mix water demand:

$$W/C = 0.26, \ W = C \cdot (W/C) = 492 \cdot 0.26 = 128 \ \text{l/m}^3$$

Table 6.8: Optimisation results for mathematical models obtained by using the MS Excel 'Solver' tool

Indicator	Units	Value
Optimal Values of Factors		
Cement consumption per 1 m³	kg	492
Slag consumption per 1 m³	kg	200
SP content	%	1
Concrete Quality Indicators		
Normative compressive strength at one day	MPa	19.3
Calculated* compressive strength at one day	MPa	21.1
Normative value of modulus of elasticity at one day	MPa	23000
Calculated* value of modulus of elasticity at one day	MPa	23000
Normative compressive strength at 28 days	MPa	77.1
Calculated* compressive strength at 28 days`	MPa	80.9

* The values of indicators calculated according to mathematical models for corresponding optimal values of factors (cement, slag and *SP* contents)

3. Find the crushed stone consumption from the condition that its volume should be 0.35 of concrete volume:

$$CS = 0.35 \cdot \rho_{CS} = 0.35 \cdot 2700 = 945 \text{ kg/m}^3$$

4. Find the sand consumption:

$$S = \left(0.65 - \frac{C}{\rho_C} - \frac{W}{\rho_W} - \frac{Slag}{\rho_{slag}} \right) \cdot \rho_{CS}$$

$$= \left(0.65 - \frac{492}{3100} - \frac{128}{1000} - \frac{200}{2900} \right) \cdot 2650 = 780 \text{ kg/m}^3$$

5. Find the superplasticizer content:

$$B_{SP} = \frac{C \cdot SP}{100} = \frac{492.1}{100} = 4.92 \text{ kg}$$

Concrete composition per 1 m³ is:

Cement – 492 kg; Water – 128 l; SP – 4.92 kg; Slag – 200 kg; sand – 780 kg; Crushed stone – 945 kg.

When using raw materials that differ from those used in the obtaining models (Table 6.7), it is effective to use the method of experimental-calculation adaptation [50], allowing to adjust the values of models' coefficients considering laboratory control data of the required indicators.

Example 3. *Determine with the help of experimental-statistical models the compositions of hydraulic concrete with specified values of compressive strength, frost resistance and water impermeability.*

When performing experiments with the help of mathematical planning methods, Portland cement, granite crushed stone, quartz sand with $M_f = 2.1$ and superplasticizer of polycarboxylate type Dynamon SP-3 were used. The conditions for planning experiments are given in Table 9. The complex of experimental and statistical models obtained during statistical processing of the results of the experiments is given in Table 10.

Table 6.9: Conditions for planning experiments

Factors		Levels of variation			Variation intervals
Natural	*Coded*	*−1*	*0*	*+1*	
Water content of concrete mixtures, kg/m³	X_1	110	130	150	20
Cement-water ratio	X_2	2.0	3.0	4.0	1.0
Maximum size of crushed stone, mm	X_3	10	40	70	30
Consumption of air entrained admixture (AEAd – Vinsol resin), kg/m³	X_4	0	0.06	0.12	0.06
Cement normal consistency, %	X_5	24.6	27.2	29.8	2.6
Cement strength, MPa	X_6	34.5	41.2	47.9	6.7
Conventional indicator of workability*	X_7	0	1	2	1
Duration of normal hardening, days	X_8	lg28	lg71	lg180	0.40

* The workability of the concrete mixture is found on the scale:

Conventional indicator	0	0.6	1	1.4	1.8	2
		Vebe, s		Slump, mm		
	40 s	2 s	5 s	80 mm	110 mm	130 mm

The resulting complex of models allows solving a number of problems:

1. In a wide range of compositions of concrete mixtures, predict the strength, frost resistance and water impermeability of concrete.
2. To find the values of frost resistance and water impermeability of concrete with known strength values and corresponding composition parameters. Similarly, it is possible to find the value of strength with known values of frost resistance and water impermeability.
3. To find compositions of concrete mixtures that correspond to the given indicators of strength, frost resistance and water impermeability of concrete.

Table 6.10: Polynomial models of properties of concrete mixture and concrete

Property	Regression equation
Water consumption of the mixture, $1/m^3$	$y_1 = 130 + 19x_7 + 9.7x_2 - 14.4x_3 - 6x_4 + + 6.7x_5 -$ $3.7x_7^2 + 4.1x_2^2 + 6.3x_3^2 + 2x_4^2 + + 5.5x_5^2 + 0.7x_7x_3 + 1.9x_8x_5 -$ $0.7x_2x_3 + + 2.2x_2x_4 + 1.6x_2x_5 - 0.7x_3x_5 + 0.7x_4x_5 \qquad (6.45)$
The optimal proportion of sand in the mixture of aggregates	$y_2 = 0.284 + 0.03x_1 - 0.039x_2 - 0.02x_3 + + 0.009x_4 + 0.007x_1^2 +$ $0.016x_2^2 + 0.08x_3^2 + + 0.006x_4^2 - 0.005x_1x_2 + 0.01x_1x_3 +$ $+ 0.09x_2x_4 - 0.004x_3x_4 + 0.007x_4x_3 \qquad (6.46)$
Compressive strength, MPa	$y_3 = 53.2 - 2.7x_7 + 21.2x_2 - 5.2x_4$ $-1.24x_5 + 7.2x_6 + 8.8x_8 - 0.07x_7^2 - 2.7x_2^2$ $-0.3x_4^2 - 0.43x_5^2 + 0.72x_6^2 - 1.15x_8^2$ $-0.9x_7x_4 - 0.65x_7x_5 + 1.8x_7x_6 + 1.4x_2x_4$ $-0.9x_2x_5 + 3.1x_2x_6 + 3.1x_2x_7 + 1.4x_4x_6$ $-0.7x_5x_6 + 1.27x_6x_7 \qquad (6.47)$
Frost resistance, cycles	$y_4 = 520.3 - 67.8x_1 + 162.3x_2 + 147.7x_4 - - 27.4x_5 + 21.8x_6 +$ $63.2x_8 + 3.7x_1^2 - - 38.8x_7^2 + 6.7x_4^2 - 9.8x_5^2 + 11.7x_6^2 + 2.2x_8^2$ $+ 29.4x_7x_2 + 15.2x_7x_4 - 8.7x_7x_5 - 16.6x_7x_8 + + 26.6x_2x_4 -$ $15.8x_2x_5 + 18.7x_2x_8 + + 7.1x_4x_6 + 23.7x_4x_8 + 13.8x_6x_8 \qquad (6.48)$
Water impermeability, MPa$\times 10^{-1}$	$y_5 = 10.53 + 0.04x_7 + 5.84x_2 + 0.02x_4 + 0.1x_5 + 0.06x_6^2 +$ $0.25x_8^2 - 0.05x_7^2 + + 2.8x_2^2 - 0.01x_4^2 + 0.004x_5^2 + 0.02x_7x_2$ $+ 0.01.x_7x_4 - 0.02_x7_x5 + 0.04x_2x_6 + 0.14x_2x_8 \qquad (6.49)$

For example, if it is necessary to calculate the compositions of concrete at a given strength at a certain age and to determine their frost resistance and water impermeability, the calculations are carried out in the following sequence:

1. Determine at the given indicators of the factors (Table 6.10) in coded form (x_8–workability of concrete mixture, x_4 – consumption of AEAd, x_3 – normal consistency of cement, x_6 – cement strength, x_9 – duration of hardening) from equation y_2 value of C/W (x_2) for given strength.
2. Next, the indicators of frost resistance and water impermeability are calculated according to models y_4 and y_5.
3. With the calculated value of C/W (x_2) and the specified values of other factors, find the water consumption (W) and the optimal proportion of sand (r) in the aggregate mixture.

4. With the determined C/W, W and r, as well as the AEAd admixture content, calculate the composition of the concrete mixture.

If the indicators of strength, frost resistance and water impermeability are given, to calculate the composition of concrete, determine C/W (x_2), which provides all three indicators. To narrow the area of C/W, the expediency of using an air-entraining admixture is being considered.

Examples of calculated compositions of concrete with design indicators of strength, frost resistance and water impermeability are given in Table 6.11.

Table 6.11: Calculated compositions of concrete

No.	Properties of concrete			Calculated parameters of concrete compositions			Concrete composition, kg/m³			
	Level of concrete according to frost resistance	*Class of concrete according to compressive strength*	*Concrete level for water impermeability*	*C/W*	*W, l/ m³*	*r*	*Cement, C*	*Water, W*	*Sand, S*	*Crushed stone, CS*
Concrete without Air-entraining Admixture										
1	F250	C 32/40	P6	2.32	123	0.41	285	123	853	1250
2	F250	C 35/45	P8	2.61	124	0.37	324	124	757	1312
3	F300	C 40/50	P10	2.91	125	0.33	364	125	663	1371
4	F350	C 45/55	P12	3.24	128	0.29	415	128	567	1415
5	F400	C 50/60	P14	3.61	132	0.25	477	132	473	1446
Concrete with Air-entraining Admixture (0.06 kg/m³)										
1	F400	C 32/40	P8	2.57	126.6	0.34	325	127	692	1369
2	F450	C 35/45	P8	2.85	128.6	0.33	367	129	659	1362
3	F500	C 40/50	P10	3.15	131.5	0.32	414	132	623	1349
4	F550	C 45/55	P14	3.48	135.6	0.31	472	136	585	1327
5	F600	C 50/60	P16	3.84	141.0	0.30	541	141	544	1293

Note: The concrete compositions are calculated to obtain a concrete mixture with a slump 50 mm when using the following materials: Portland cement with an strength of 47.2 MPa and a normal consistency of 27.2%, as well as crushed stone with a maximum size of 40 mm. The specified duration of concrete hardening is 70 days.

Part II

High-strength
Fibre-reinforced Concrete

Part II
High-strength
Fibre-reinforced Concrete

The Influence of Dispersed Reinforcement Factors on the Strength Properties of Fibre-reinforced Concrete

Effect of Fibre Geometric Characteristics

There are quite many types of available steel fibres, which differ in geometric characteristics and ways of anchoring in the concrete matrix. Research data on the effect of fibre type, shape, method of anchoring on steel fibre-reinforced concrete strength indicators are quite contradictory. Some results demonstrate the advantage of steel fibre with anchors at the ends (hooked ends) [51, 52], while others, with corrugated shape [53]. Common in all cases is the importance of ensuring high bond strength of dispersed reinforcement with concrete.

The present research is focused on investigating the influence of composition parameters at using various available types of fibre on producing high-strength steel fibre-reinforced concrete with rather low consumptions of cement and fibre. The following types of fibres were used in the research (Fig. 7.1):

1. corrugated type Fibax F1 60/1;
2. anchored with hooked ends 'Dramix' produced by 'Becaert Ltd.' F2 60/1;
3. anchor with flattened ends 'Mixarm' F3 50/1;
4. anchor with bent ends 'Chelyabinka' F4 33/0.85/0.75;
5. straight fibre with anchors in a form of cones 'Mixarm' F5 54/1.

The main characteristics of these fibre types are given in Table 7.1.

At the first stage, research was carried out by using three types of fibre: F1, F2 and F3. To determine the fibre type that allows to obtain concrete withthe highest flexural strength, experiments were conducted using mathematical planning [42].

F2   F1

F3

F4

F5   d

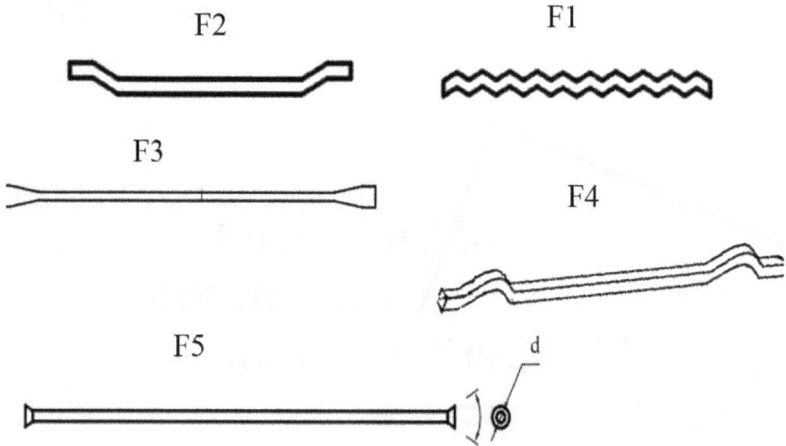

Fig. 7.1: Main types of fibre

Table 7.1: Types of fibre: Main characteristics

Main indicators	Fibre type				
	F1 60/1	*F2 60/1*	*F3 50/1*	*F4 33/0.85/ 0.75*	*F5 54/1*
Length (L), mm	60.0±6.0	60.0 ± 6.0	50 ± 5.0	33.0 ± 3.0	54.0 ± 4.0
Diameter (d), mm	1.0 ± 0.1	1.0 ± 0.1	1.0 ± 0.1	0.9 ± 0.1	1.0 ± 0.03
$\Lambda = L/d$	60	60	50	37	54
Tensile strength, MPa at least	1335	1335	1335	1260	1100
Length of hooked/ flattened end, mm	-	5.0 ± 0.1	4.0 ± 0.1	2.5 ± 0.1	2.0 ± 0.1
Wave/hooked end height, mm	4.5 ± 0.1	5.0 ± 0.1	-	5.0 ± 0.1	-
Average steel density (ρ), g/cm^3	7.86	7.86	7.86	7.86	7.86

Two semi-replicas of type 2^{3-1} were realised. The experiments' planning conditions are given in Table 7.2. Research was carried out, using two types of concrete: normal-weight concrete with crushed stone fraction of 5-20 mm as a coarse aggregate and fine-grained concrete with a mixture of 0.16-2 mm quartz sand and 2-5 mm crushed granite stone.

As initial components of the concrete mixture used were cement CEM I M500, quartz sand with $M_f = 2.1$, crushed granite stone fraction 5-20 mm, fibre content was 80 kg/m^3 ($\mu = 1.0\%$). Melflux 2651F polycarboxylate type superplasticizer admixture was added to concrete mixture. The ratio of sand and crushed stone for normal-weight concrete was calculated according to known recommendations [24]. Concrete mixtures had the same workability ($Sl = 150$ mm).

Table 7.2: Experiments planning conditions for choosing the fibre type

Influence factors		Factor variation levels	
Natural form	*Coded form*	*−1*	*+1*
Fibre type	X_1	Anchor - type*	Corrugated
W/C	X_2	0.35	0.45
Cement consumption, kg/m³	X_3	500	600

* In the first semi-replica, an anchor type fibre with hooked ends (F2) was used, and in the second, one with flattened ends (F3)

Previously, experiments were conducted for two types of concrete without fibre. The values of compressive and flexural strengths for used concrete are given in Table 7.3. As can be seen from the given data, the influence of cement consumption and water-cement ratio on strength values is expected. Increasing W/C from 0.35 to 0.45 reduces the compressive strength by 20-25% and the flexural strength to a lesser extent. An increase in cement consumption at the same W/C has little effect on the strength values.

Table 7.3: Concrete strength vs. cement consumption and W/C

W/C	Cement consumption, kg/m³	Flexural strength at seven days, MPa	Compressive strength at seven days, Mpa	Flexural strength at 28 days, MPa	Compressive strength at 28 days, MPa
Normal-weight Concrete					
0.35	500	3.1	75.8	3.7	80.3
0.35	600	3.0	69.5	3.8	82.1
0.45	500	2.5	63.8	3.2	68.1
0.45	600	2.9	65.3	3.1	73.3
Fine-grained Concrete					
0.35	500	3.1	61.1	4.0	73.9
0.35	600	4.2	64.3	4.3	76.3
0.45	500	3.5	43.6	3.6	58.8
0.45	600	3.0	48.1	3.8	63.9

As a result of the two semi-replicas realisation and statistical processing of experimental data, polynomial models of the following form were obtained:

$$y = b_0 + b_1X_1 + b_2X_2 + b_3X_3 + b_{12}X_1X_2 \qquad (7.1)$$

Coefficients of mathematical models for compressive and flexural strength of normal-weight concrete are given in Table 7.4.

Table 7.4: Coefficients of mathematical strength models and comparison of fibre types for normal-weight concrete

Output parameters	Values of coefficients*				
	b_0	b_1	b_2	b_3	b_{12}
Flexural strength at one day, Mpa	3.26/ 2.98	-1.37/ -1.21	-0.2/ -0.1	0.05/ 0.03	2.4/ 0.56
Compressive strength at one day, Mpa	38.2/ 39.1	-1.5/ -2.3	-3.6/ -2.1	0.9/ 2.5	0.3/ 1.1
Flexural strength at seven days, Mpa	4.55/ 4.48	-1.25/ -1.33	-0.1/ -0.2	0.1/ 0.03	0.1/ 0.03
Flexural strength at 28 days, Mpa	6.0/ 5.78	-1.4/ -1.63	-0.3/ -0.3	0.2/ 0.18	0.2/ 0.18
Compressive strength at seven days, Mpa	62.3/ 64.3	-2.25/ -0.25	-6.8/ -3.8	-2.3/ 0.75	-2.3/ 0.75
Compressive strength at 28 days, MPa	77.3/ 74.8	-3.5/ -6	-5.3/ -4.3	0.5/ 1.5	0.5/ 1.5

* In the numerator are shown coefficients of the model for comparison of F1 and F2 fibre types; coefficients of the model for comparison fibre types F1 and F3 are given in the denominator

Figures 7.2-7.4 present graphical dependencies obtained by corresponding calculations, using the models characterising the dependence of concrete flexural and compressive strengths ($f_{c,tf}$ and f_{cm}) at one, seven and 28 days vs. W/C for different types of steel fibre.

Analysis of mathematical models and graphical dependencies demonstrates that the highest compressive and especially flexural strength values are achieved when using corrugated fibres (F1). This, obviously, can be explained by the increased adhesion surface of such a fibre with the concrete in comparison with the anchor type fibre.

An increase in the water-cement ratio leads to a decrease in strength in all terms and to a higher extent in compressive strength (by 20-30%). An increase in cement consumption at a constant W/C leads to higher strength, but the influence of this factor within the factors variation range is insignificant.

Coefficients of mathematical models for compressive and flexural strength of fine-grained concrete are given in Table 7.5. Figures 7.5-7.7 present graphical dependencies obtained by corresponding calculations based on the models characterising the dependence of the flexural and compressive strengths of fine-

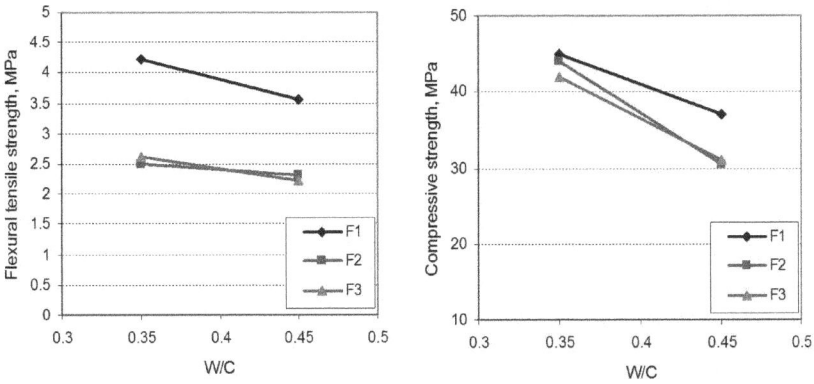

Fig. 7.2: Calculated dependencies of flexural and compressive strengths of steel fibre-reinforced concrete at one day vs. *W/C*

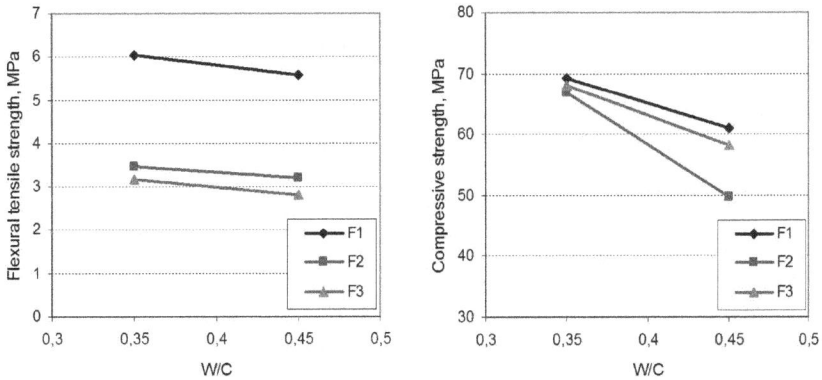

Fig. 7.3: Calculated dependencies of flexural and compressive strengths of steel fibre-reinforced concrete at seven days vs. *W/C*

grained concrete ($f_{c,tf}$ and f_{cm}) at one, seven and 28 days vs. *W/C* for different types of steel fibre.

Analysis of mathematical models (Table 7.5) and graphic dependencies (Figs. 7.5-7.7) demonstrate that the fine-grained fibre-reinforced concrete strength at all hardening stages depends on the influence factors in a similar way as for normal-weight concrete. At the same time, the absolute values of flexural strength are significantly higher as compared to normal-weight concrete (by 40-50%), which can be explained by the increase in the contact surface area between the fibre and fine-grained concrete mortar matrix. The compressive strength increases insignificantly within 5-10%, depending on the fibre type and the specimens age.

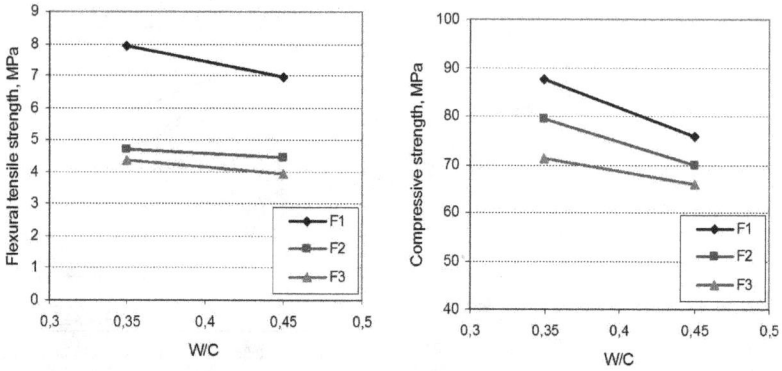

Fig. 7.4: Calculated dependencies of flexural and compressive strengths of steel fibre-reinforced concrete at 28 days vs. W/C

Table 7.5: Coefficients of mathematical strength models and comparison of fibre types for fine-grained concrete

Output parameters	Values of coefficients*				
	b_0	b_1	b_2	b_3	b_{12}
Flexural strength at one day, Mpa	5.3/ 5.84	-1.65/ -2.05	-1.2/ -1.3	0.6/ 0.22	1.6/ 0.18
Compressive strength at one day, Mpa	40.3/ 41.7	-1.3/ -1.8	-4.3/ -3.3	0.4/ 0.8	0.5/ 1.2
Flexural strength at seven days, Mpa	7.32/ 6.8	-2.3/ -1.7	-1.1/ -1.3	0.15/ 0.2	0.15/ 0.9
Flexural strength at 28 days, Mpa	10.2/ 11.3	-2.1/ -0.8	-0.9/ -1.2	0.3/ 0.4	0.26/ -0.3
Compressive strength at seven days, MPa	64.1/ 64.9	-1.35/ -0.25	-5.7/ -4.1	1.2/ 0.95	1.1/ 0.9
Compressive strength at 28 days, Mpa	80.1/ 77.8	-2.2/ -5.1	-4.2/ -3.2	0.9/ 1.2	3.5/ 0.8

* In the numerator are shown coefficients of the model for comparison of F1 and F2 fibre types; coefficients of the model for comparison of fibre types F1 and F3 are given in the denominator.

The highest strength values are also achieved for fibre type F1. However, this dependence is most clearly manifested when determining the flexural strength, while for the compressive strength at low W/C values, there is even a slight predominance of fibre type F2 (Fig. 7.7). An increase in water-cement ratio leads to a decrease in strength at all hardening stages, and to a higher extent for compressive strength.

Considering the kinetics of strength growth for fibre-reinforced concrete over time (Figs. 7.8 and 7.9), it can be noted that its value at one day is about 50% of the 28-day value, and at seven days, it is about 80%.

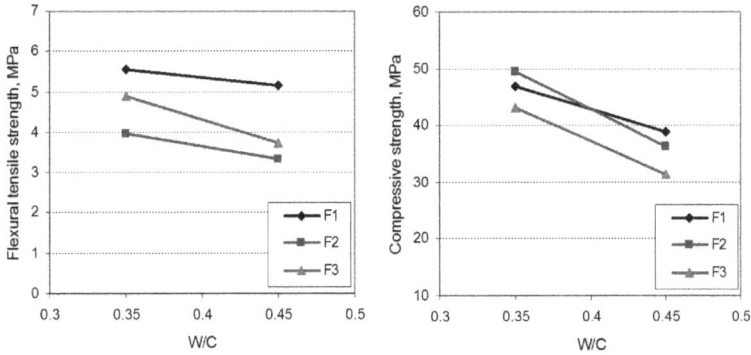

Fig. 7.5: Calculated dependencies of flexural and compressive strengths for steel fibre-reinforced fine-grained concrete at one day vs. *W/C*

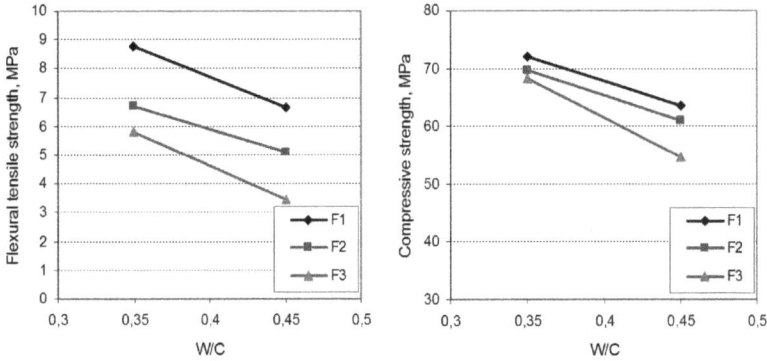

Fig. 7.6: Calculated dependencies of flexural and compressive strengths for steel fibre-reinforced fine-grained concrete at seven days vs. *W/C*

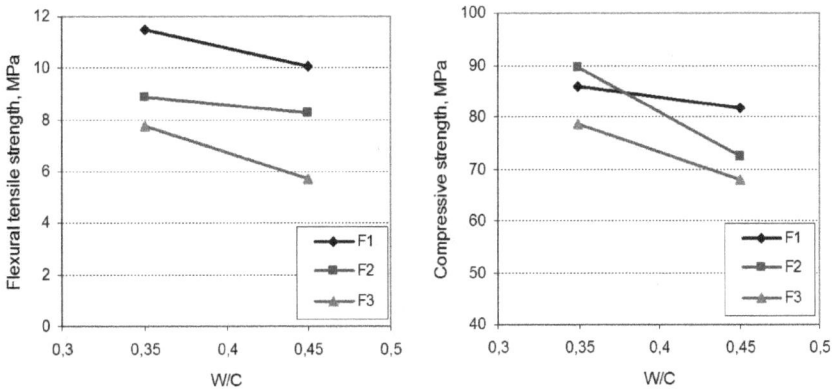

Fig. 7.7: Calculated dependencies of flexural and compressive strengths for steel fibre-reinforced fine-grained concrete at 28 days vs. *W/C*

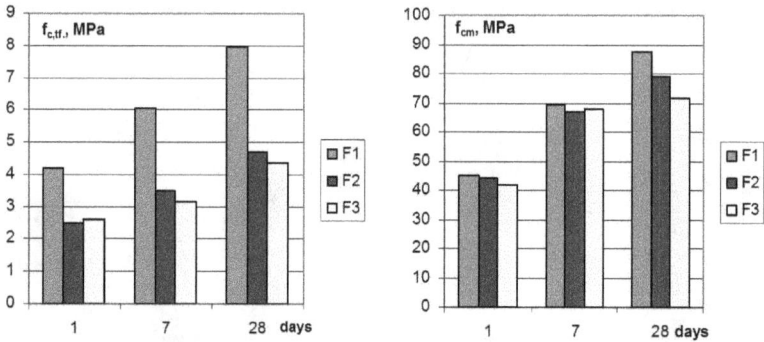

Fig. 7.8: Kinetics of flexural and compressive strengths growth for steel fibre-reinforced normal-weight concrete with different types of fibre

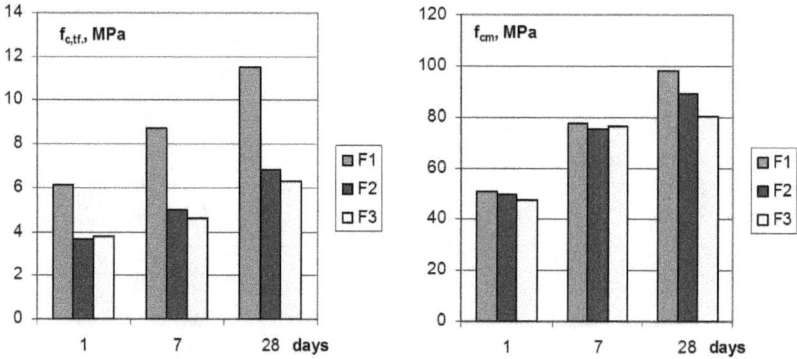

Fig. 7.9: Kinetics of flexural and compressive strengths growth for steel fibre-reinforced fine-grained concrete with different types of fibre

The first stage of the study showed higher efficiency of fibre type F1. To compare the effectiveness of dispersed reinforcement with two others types of fibre, namely F4 and F5, a number of additional experiments were conducted at the second stage. Steel fibre-reinforced concrete specimens were made from mixtures with $W/C = 0.35$ and with a cement consumption of 500 kg/m^3.

Diagrams for comparison of strength of specimens for all types of fibre and without it are shown in Figs. 7.10 and 7.11.

As can be seen from the obtained results, fibre type F4 has shown a significant increase in flexural strength, compared to the base composition, especially for fine-grained concrete, but at the same time it is significantly inferior compared to fibre type F1. Fibre type F5 showed results similar to those obtained for fibre types F3 and F4.

Obviously, the main effect of dispersed reinforcement in concrete is manifested in an increase in the $f_{c,tf}/f_{cm}$ ratio (efficiency coefficient). This can

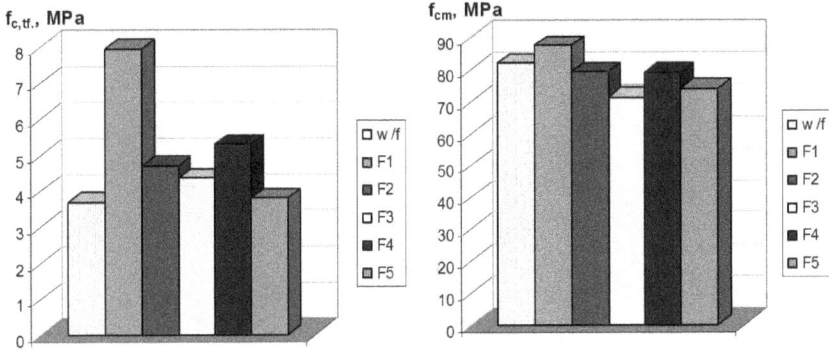

Fig. 7.10: The influence of fibre type on reinforced normal-weight concrete flexural and compressive strengths at 28 days ($\mu = 0.5\%$); * w/f – without fibre

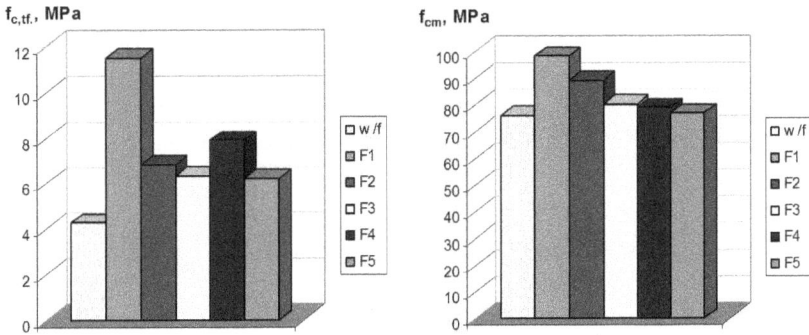

Fig. 7.11: The influence of fibre type on fine-grained reinforced concrete flexural and compressive strengths at 28 days ($\mu = 0.5\%$);* w/f – without fibre

be clearly seen in Fig. 7.12, which shows the relative efficiency coefficients of different fibre types as dispersed reinforcement for fine-grained concrete.

As a reference (100%) was selected the $f_{c,tf}/f_{cm}$ ratio for fine-grained fibre concrete with fibre type F1. For all types of fibre, the efficiency factor is higher than for non-fibre-reinforced concrete.

Thus, steel corrugated fibre, which is most effective in terms of increasing flexural strength, was chosen for further research.

Selection of the Optimal Steel Fibre Content

With an increase in the dispersed reinforcement volume concentration, the physical and mechanical characteristics of fibre-reinforced concrete increase to a certain level, after reaching which these characteristics begin to decrease [54]. Therefore, there is a certain ratio of fibre and matrix volumes, at which the physical and mechanical characteristics of fibre-reinforced concrete will have a maximum value.

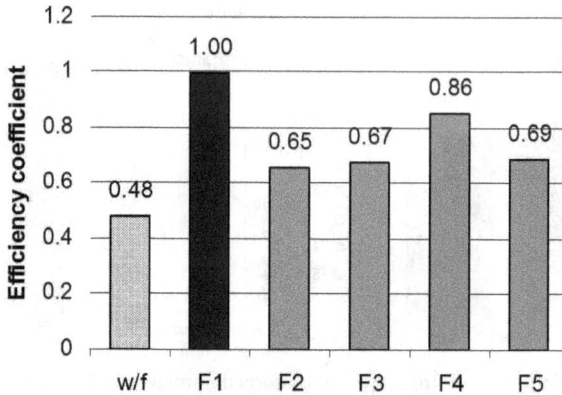

Fig. 7.12: Relative values of the dispersed reinforcement efficiency coefficients for different types of fibres of fine-grained concrete

It was demonstrated [55] that within the interval of the minimum and maximum reinforcement ratio ($\mu_{min} - \mu_{max}$), there is a characteristic point corresponding to the moment of fibre cement frame formation (μ_k), before and after which the behaviour of the concrete and its properties differ significantly (Fig. 7.13).

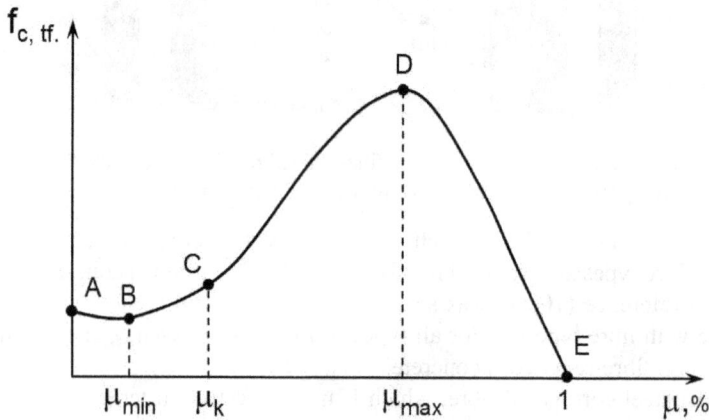

Fig. 7.13: The nature of changes in fibre-reinforced concrete strength depending on the fibres content

Interval AB characterises small saturations, when the fibres are separated from each other by significant distances ('diffuse reinforcement zone'), the strength of fibre-reinforced concrete is characterised by the strength of the matrix and practically does not differ from it. Interval BC characterises the 'zone of concentrated reinforcement', when the matrix cracks, the fibres are able to take the load and provide the load-bearing capacity of fibre-reinforced concrete.

Point C is the fusion moment of the 'fibre-matrix' contact zones and fibre-cement framework formation. In the CD interval, there is a further and more intensive increase in the fibre-reinforced concrete strength, which is a result of cement-stone compaction between the fibres. Point D corresponds to the maximum fibre-reinforced concrete strength, a further decrease of which is due to a decrease in the thickness of the matrix layer to such an extent that the material shows a tendency to delaminate even under small loads [55].

Table 7.6: Experiment planning conditions for determining parameters of steel fibre-reinforced concrete compositions

Number of factors	Factors		Variation levels			Interval
	Code	Natural form	−1	0	+1	
1	X_1	Cement consumption, kg/m³ (C)	450	500	550	50
2	X_2	W/C	0.35	0.4	0.45	0.1
3	X_3	Fibre content, kg/m³ (F)	0	40	80	40

The experimental results are shown in Table 7.

Table 7.7: Experimental results determination of the optimal fibre content

Plan point number	Values of coded factors			Components consumptions, kg/m³					W/C	SP, %	Sl, sm	f_{cm}^3 MPa	$f_{c,tf}^3$ MPa	f_{cm}^7 MPa	$f_{c,tf}^7$ MPa	f_{cm}^{28} MPa	$f_{c,tf}^{28}$ MPa
	X_1	X_2	X_3	C	S	A	W	F									
1	+	+	+	550	419	1206	247	40	0.45	0.25	18	26.5	2.6	37.5	3.5	58	5.6
2	+	+	-	550	419	1206	247	0	0.45	0.14	19	25.5	1.1	35	1.8	54	2.2
3	+	-	+	550	580	1206	193	40	0.35	0.92	19	55.3	4.3	60	5.6	86	6.9
4	+	-	-	550	580	1206	193	0	0.35	0.68	18	61.1	2.5	69	3.1	82	3.1
5	-	+	+	450	646	1206	203	40	0.45	0.41	18.5	41.2	3.4	54.5	4.6	62	5.8
6	-	+	-	450	646	1206	203	0	0.45	0.26	16	44.8	2.2	55.6	2.7	69	2.6
7	-	-	+	450	779	1206	157	40	0.35	1.13	19.5	42.5	1.8	66.8	5	80	6.6
8	-	-	-	450	779	1206	157	0	0.35	0.84	18.5	41	1.9	61	2.1	72	2.2
9	+	0	0	550	499	1206	220	20	0.4	0.40	17	37	2.9	52	3.3	60	3.8
10	-	0	0	450	714	1206	180	20	0.4	0.56	18	38.5	3.1	54.6	3.9	66	5.4
11	0	+	0	500	531	1206	226	20	0.45	0.03	18	27.5	2.2	40.2	2.9	45	3.8
12	0	-	0	500	679	1206	176	20	0.35	0.66	18	42.6	3.2	50	3.8	60	4.9
13	0	0	+	500	606	1206	200	40	0,4	0.27	19.5	37	3.5	47	4.3	57	5.5
14	0	0	-	500	606	1206	200	0	0.4	0,07	17.5	31.5	1.6	41	1.8	46	1.9
15	0	0	0	500	606	1206	200	20	0.4	0.21	17.5	37.8	2.6	48	3.3	55	4
16	0	0	0	500	606	1206	200	20	0.4	0.21	17.5	37.8	2.6	48	3.3	55	4
17	0	0	0	500	606	1206	200	20	0.4	0.21	17.5	37.8	2.6	48	3.3	55	4

At the first stage, a three-level, three-factor, close to D-optimal plan was implemented [42]. The experiment planning conditions are given in Table 7.6. Research was carried out for normal-weight concrete with the use of crushed stone fraction of 5-20 mm as a coarse aggregate and corrugated fibre F1. Melflux 2651f superplasticizer was added to the mixture. The output parameters were compressive strength f_{cm} and flexural strength $f_{c.tf}$ at three, seven and 28 days of hardening (Table 7.7).

Statistical analysis of the obtained experimental results allowed to calculate mathematical models of fibre-reinforced concrete strength parameters (Table 7.8)

Table 7.8: Mathematical models of steel fibre-reinforced concrete strength parameters

Parameters	Mathematic models	
Admixture Melflux consumption, %	$SP = 0.21 - 0.081X_1 - 0.315X_2 + 0.099X_3 + 0.27X_1^2 + 0.14X_2^2 - 0.04X_3^2 + 0.012X_1X_2 - 0.012X_1X_3 - 0.34X_2X_3$	(7.2)
Compressive strength at hardening time: Three days	$f_{cm}^3 = 35.46 - 0.26X_1 - 7.7X_2 - 0.14X_3 + 4.171X_1^2 + 1.47X_2^2 + 0.671X_3^2\,2 - 8.363X_1X_2 - 0.388X_1X_3 + 0.213X_2X_3$	(7.3)
Seven days	$f_{cm}^7 = 46.20 - 3.9X_1 - 8.4X_2 + 0.42X_3 + 8.618X_1^2 + 0.418X_2^2 - 0.682X_3^2 - 4.850X_1X_2 - 1.40X_1X_3 + 0.575X_2X_3$	(7.4)
28 days	$f_{cm}^{28} = 60.6 + 0.72X_1 - 16.5X_2 - 4.8X_3 + 11.748X_1^2 + 2.29X_2^2 + 0.498X_3^2 - 6.70X_1X_2 + 1.80X_1X_3 - 2.70X_2X_3$	(7.5)
Flexural strength at hardening time: Three days	$f_{c.tf}^3 = 2.74 + 0.1X_1 - 0.22X_2 + 0.63X_2X_3 + 0.171X_1^2 - 0.129X_2^2 - 0.279X_3^2 - 0.625X_1X_2 + 0.275X_1X_3 + 0.12X_2X_3$	(7.6)
Seven days	$f_{c.tf}^7 = 3.27 - 0.1X_1 - 0.41X_2 + 1.15X_3 + 0.365X_1^2 - 0.106X_2^2 - 0.194X_3^2 - 0.450X_1X_2 - 0.075X_1X_3 - 0.225X_2X_3$	(7.7)
28 days	$f_{c.tf}^{28} = 4.05 - 0.15X_1 - 0.44X_2 + 1.67X_3 + 0.532X_1^2 - 0.282X_2^2 - 0.368X_3^2 - 0.388X_1X_2 - 0.087X_1X_3 - 0.213X_2X_3$	(7.8)

and to obtain graphic dependencies based on them (Figs. 7.14-7.17). In order to obtain graphical dependencies for two factors, the value of the third one was assumed to remain at the basic (zero) level.

Analysis of models and graphical dependencies confirm the most significant role of dispersed reinforcement on the value of flexural strength. Increasing the fibre content from 0 to 80 kg/m^3 leads to an increase in flexural strength at 28 days by more than two times. The remaining technological factors practically do not affect $f_{c,tf}$. Regarding compressive strength, the water-cement ratio factor is expected to have the highest influence. A change in fibre content within the investigated limits leads to a slight increase in compressive strength and this is observed at low W/C values and cement consumption.

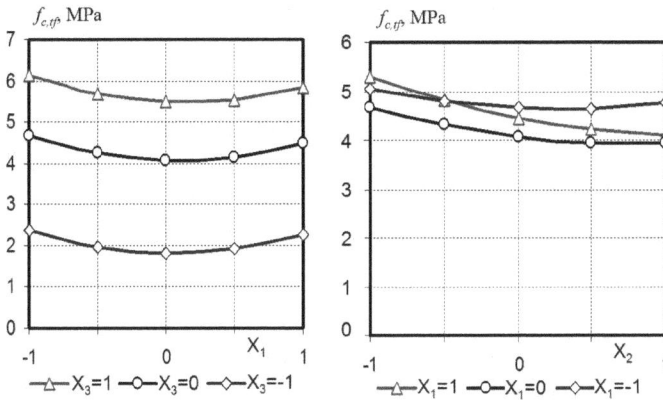

Fig. 7.14: Dependencies of steel fibre-reinforced concrete flexural strength at 28 days on cement consumption (X_1), W/C (X_2) and fibre content (X_3)

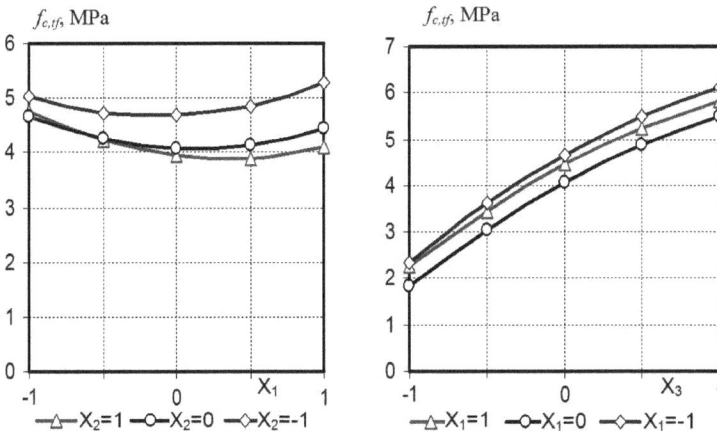

Fig. 7.15: Dependencies of steel fibre-reinforced concrete flexural strength at 28 days on cement consumption (X_1), W/C (X_2) and fibre content (X_3)

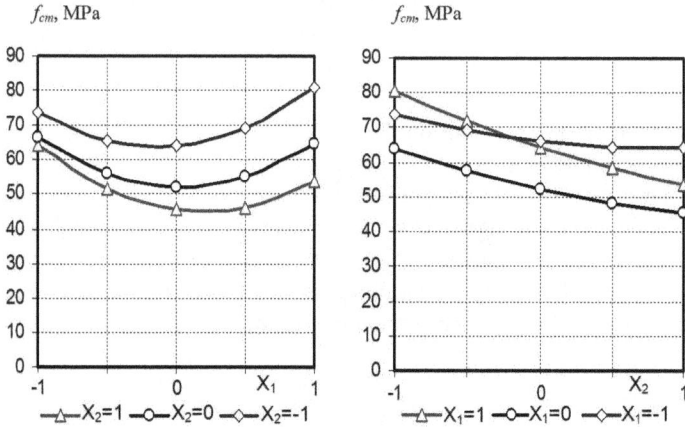

Fig. 7.16: Dependencies of steel fibre-reinforced concrete compressive strength at 28 days on cement consumption (X_1) and W/C (X_2)

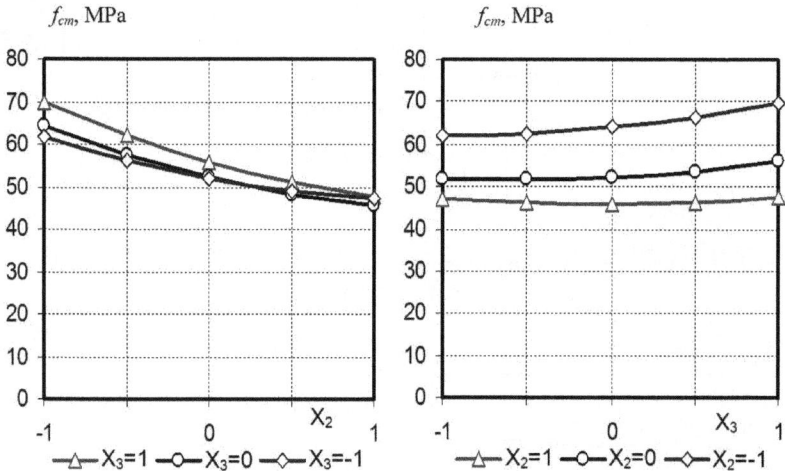

Fig. 7.17: Dependencies of steel fibre-reinforced concrete compressive strength at 28 days on W/C (X_2) and fibre content (X_3)

The relative influence of factors on the compressive strength and flexural strength values can be followed from Fig. 7.18, whereas a unit value was assumed to influence (linear coefficient in regression equations) factor X_3 (fibre content).

The next research stage was focused on establishing the optimal steel fibre content for production of fine-grained steel fibre-reinforced concrete with high flexural strength. The research was conducted using fibre type F1, as the increase in its content, according to the results of previous studies, leads to intensive increase in concrete flexural strength.

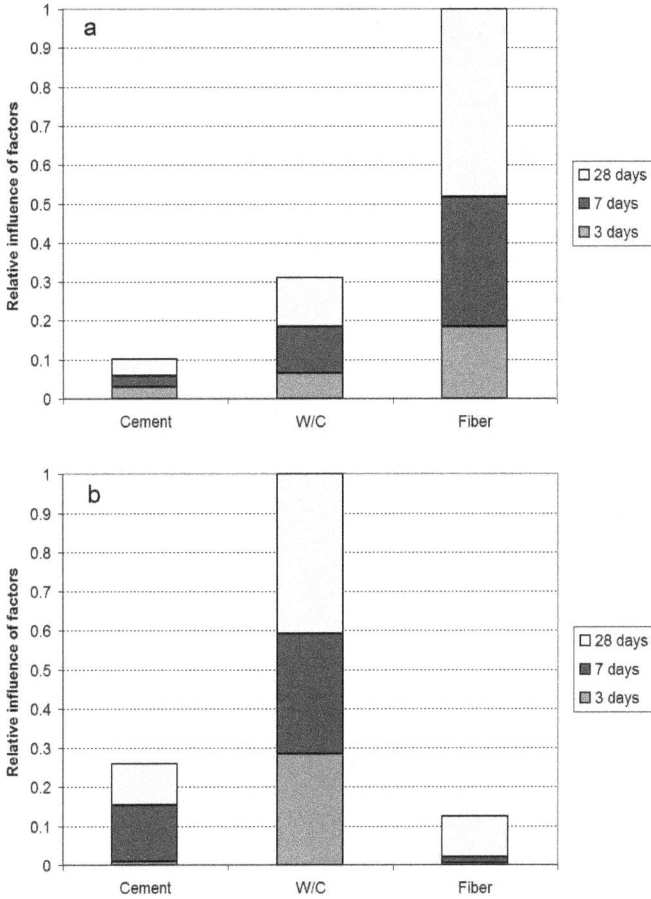

Fig. 7.18: Relative influence of factors on the flexural strength (a) and compressive strength (b) of steel fibre-reinforced concrete

The aim of the work at this research stage was to determine the complex effect of cement consumption, fibre and water-cement ratio on fine-grained steel fibre-reinforced concrete strength characteristics. For this purpose, a three-level, three-factor, close to D-optimal plan was implemented [42]. The experiment planning conditions and the results of the research are presented in Tables 7.9 and 7.10, respectively.

After processing and statistical analysis of the experimental data, mathematical models of fine-grained steel-reinforced concrete compressive and flexural strengths were obtained in the form of polynomial regression equations. Results of experimental data processing and statistical analysis are shown in Table 7.11 and in Figs. 7.19-7.22.

Table 7.9: Experiment planning conditions for determining the steel fibre-reinforced concrete compositions parameters

Number of factors	Factors		Variation levels			Variation interval
	Code	Natural form	-1	0	$+1$	
1	X_1	Cement consumption, kg/m³ (C)	450	500	550	50
2	X_2	W/C	0.3	0.35	0.4	0.05
3	X_3	Fibre content, kg/m³ (F)	80	100	120	20

Table 7.10: Experimental values of steel fibre-reinforced concrete compositions parameters

Plan point number	Coded values of factors			Consumptions of components, kg/m³					W/C	SP, %	f_{cm}^1, MPa	$f_{c,tf}^1$, MPa	f_{cm}^7, Mpa	$f_{c,tf}^7$, MPa	f_{cm}^{28}, MPa	$f_{c,tf}^{28}$, MPa
	X_1	X_2	X_3	C	S	A	W	F								
1	+	+	+	550	732	895	220	120	0.4	0.2	28.63	6.22	54.7	11.15	67.2	13.84
2	+	+	-	550	732	895	220	80	0.4	0.1	28.21	5.00	55.8	10.05	66.6	9.08
3	+	-	+	550	799	977	165	120	0.3	1.1	40.9	8.80	78.1	15.95	96.0	19.92
4	+	-	-	550	799	977	165	80	0.3	0.8	40.3	7.78	79.7	15.57	95.2	15.23
5	-	+	+	450	820	1002	180	120	0.4	0.4	23.8	6.64	49.1	13.91	57.3	14.90
6	-	+	-	450	820	1002	180	80	0.4	0.2	23	5.26	48.0	8.29	56.9	9.59
7	-	-	+	450	875	1069	135	120	0.3	1.3	34.1	7.50	71.0	15.27	83.9	17.06
8	-	-	-	450	875	1069	135	80	0.3	1	33.8	6.32	70.4	10.37	83.3	11.82
9	+	0	0	550	766	936	193	100	0.35	0.45	32.2	7.90	67.3	15.57	79.7	18.13
10	-	0	0	450	847	1036	158	100	0.35	0.65	30.7	7.38	64.0	14.35	75.8	16.95
11	0	+	0	500	776	948	200	100	0.4	0.3	24.5	6.72	53.4	13.32	65.3	15.49
12	0	-	0	500	837	1023	150	100	0.3	0.5	39.1	8.54	77.4	16.76	92.4	19.64
13	0	0	+	500	806	986	175	120	0.35	0.3	32.1	7.14	66.8	13.33	79.0	15.67
14	0	0	-	500	806	986	175	80	0.35	0.1	31.1	5.94	64.5	10.33	77.5	10.67
15	0	0	0	500	806	986	175	100	0.35	0.2	31.6	7.56	66.0	14.63	78.1	17.17
16	0	0	0	500	806	986	175	100	0.35	0.2	31.6	7.56	66.0	14.63	78.1	17.17
17	0	0	0	500	806	986	175	100	0.35	0.2	31.6	7.56	66.0	14.63	78.1	17.17

Table 7.11: Mathematical models of fine-grained steel-reinforced concrete strengths parameters

Output parameter		Mathematical models
Compressive strength at hardening time:	One day	$f_{cm}^1 = 31.7 + 1.4X_1 - 8X_2 + 0.3X_3 -$ $0.2X_1^2 + 0.2X_2^2 - 0.4X_1X_2$ (7.9)
	Seven days	$f_{cm}^7 = 66.5 + 3.3X_1 - 11.6X_2 + 0.1X_3 -$ $0.9X_1^2 - 1.2X_2^2 - 0.9X_3^2 - 0.4X_1X_2 -$ $0.6X_1X_3 + 0.1X_2X_3$ (7.10)
	28 days	$f_{cm}^{28} = 78.9 + 4.8X_1 - 13.7X_2 + 0.4X_3 -$ $1.6X_1^2 - 0.4X_2^2 - X_3^2 - 0.5X_1X_2 - 0.1X_2X_3$ (7.11)
Flexural strength at hardening time:	One day	$f_{c,tf}^1 = 7.56 + 0.26X_1 - 0.91X_2 + 0.6X_3 +$ $0.077X_1^2 + 0.07X_2^2 - 1.02X_3^2 -$ $0.43X_1X_2 - 0.042X_1X_3 + 0.05X_2X_3$ (7.12)
	Seven days	$f_{c,tf}^7 = 14.63 + 0.61X_1 - 1.72X_2 + 1.5X_3 +$ $0.333X_1^2 + 0.407X_2^2 - 2.8X_3^2 -$ $0.86X_1X_2 - 1.128X_1X_3 + 0.18X_2X_3$ (7.13)
	28 days	$f_{c,tf}^{28} = 17.17 + 0.589X_1 - 2.078X_2 +$ $2.5X_3 + 0.367X_1^2 + 0.394X_2^2 - 4X_3^2 -$ $0.98X_1X_2 - 0.138X_1X_3 + 0.018X_2X_3$ (7.14)

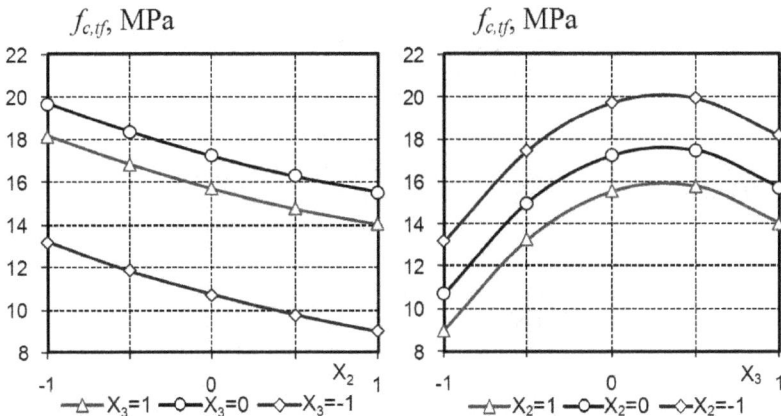

Fig. 7.19: Dependencies of fine-grained fibre-reinforced concrete flexural strength at 28 days on *W/C* (X_2) and fibre content (X_3)

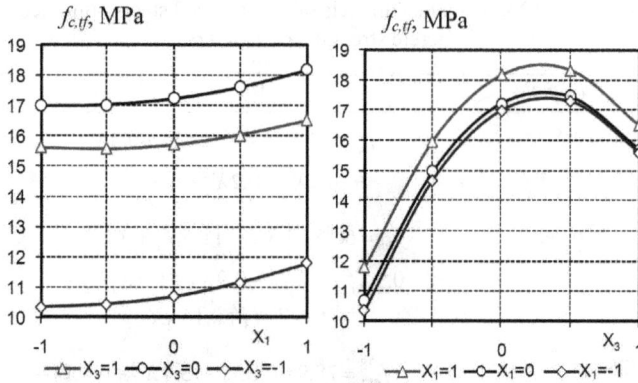

Fig. 7.20: Dependencies of fine-grained fibre-reinforced concrete flexural strength at 28 days on cement consumption (X_1) and fibre content (X_3)

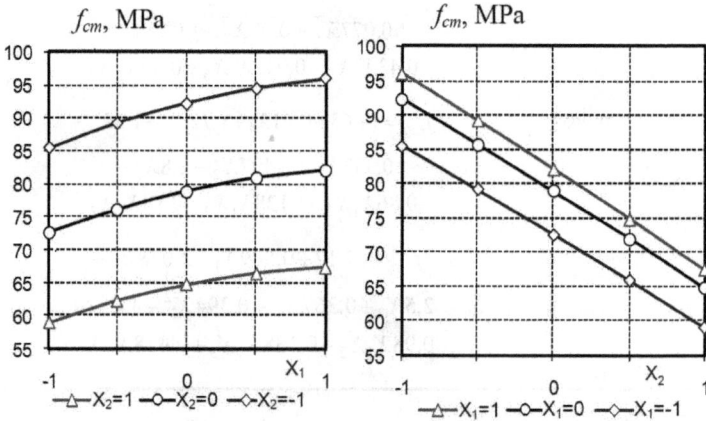

Fig. 7.21: Dependencies of fine-grained fibre-reinforced concrete compressive strength at 28 days on cement consumption (X_1) and W/C (X_2)

Analysis of the obtained mathematical models and graphical dependencies show that, as in the case of normal-weight concrete, the fibre content has the most significant effect on the flexural strength of concrete and the influence of this factor has an extreme nature. Increase in the fibre content from 80 to 100 kg/m^3 leads to an average increase in flexural strength by 40%, while a further increase in the amount of dispersed reinforcement does not significantly affect the fine-grained reinforced concrete flexural strength. That is, in accordance with the obtained experimental results, the optimal steel fibre content providing the maximum flexural strength of fibre-reinforced concrete was obtained. Further increase in the dispersed reinforcement content leads to a decrease in strength, which is due to a decrease in the matrix layer thickness to such an extent that the material shows a tendency to delamination even under small loads [54].

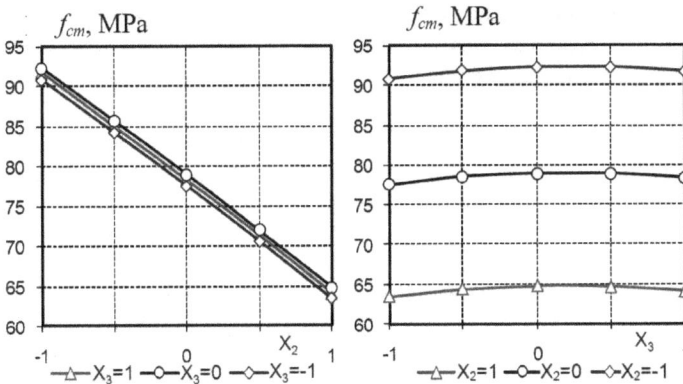

Fig. 7.22: Dependencies of fine-grained fibre-reinforced concrete compressive strength at 28 days on W/C (X_2) and fibre content (X_3)

Regarding the compressive strength, analysing the obtained experimental-statistical models, we come to the conclusion that the most significant factor affecting the fine-grained steel fibre-reinforced concrete strength at different hardening stages is the water-cement ratio, the decrease of which from $W/C =$ 0.4 to $W/C = 0.3$ leads to an increase in strength at one day by 40%, at seven days by 30% and at 28 days by 25%. The influence of this factor is linear and it accounts for about 80% of the influence of all other factors. An increase in cement consumption and steel fibre content within the varying limits at a constant water-cement ratio does not significantly affect the tested concrete specimens' strength. According to the influence of various factors on the compressive strength of fine-grained fibre-reinforced concrete, they can be arranged in the following order: $X_2 > X_1 > X_3$.

Thus, the results of the experimental programme have confirmed the possibility of obtaining fine-grained steel fibre-reinforced concrete, characterised by compressive strength of more than 90 MPa and flexural strength of almost 20 MPa at fibre content of 100 kg/m^3.

The Effect of Steel Fibres Addition Process into the Concrete Mixture

Special attention should be paid to the preparation stage of fibre-reinforced concrete mixture, because its properties and durability depend on mixing quality of the components. Several methods of obtaining steel fibre-reinforced concrete are known [56]. According to the first method, the mixture is prepared in the following sequence: first, sand is mixed with aggregate, then the necessary amount of pre-sifted fibres is added. After that, cement and water with admixtures are added to the mixture. Components are mixed until a homogeneous composition of concrete is achieved. In some cases, the fibre is added to the aggregate, pre-mixed with water and then cement and the missing amount of water are added.

The search for optimal methods of adding steel fibre into the concrete mixture led to the need for developing new proposals.

Two methods of adding steel fibres into the concrete mixture were studied (Table 7.12). According to the first method, the required amount of water was previously added to the mixed dry components of concrete, including the powdered superplasticizer . At the second stage, the necessary amount of fibre sifted through a sieve was added into the mixture prepared in this way at continuous mixing.

Table 7.12: Procedures for adding concrete mix components

Way	*Sequence of components adding*	
	First stage	*Second stage*
1. Classic	Crushed stone + sand + (cement + SP) + water	Fibre mixing
	mixing	
2. Dry	Crushed stone + sand + fibre + (cement + SP) mixing	Water mixing

According to the second method, the required amount of pre-sifted fibre was added to the mixed dry components (aggregates). Next, cement and required amount of water with plasticizing admixtures were added to the mixture and mixing was continued until a homogeneous concrete mixture was obtained.

To compare the two methods of fibre addition, experiments were conducted for normal-weight and fine-grained fibre-reinforced concrete with the optimal F1 fibre content, which was determined earlier.

For normal-weight fibre-reinforced concrete, the fibre content was 80 kg/m^3 ($\mu = 1.0\%$), for fine-grained concrete – 100 kg/m^3 ($\mu = 1.3\%$). The concrete mixtures were prepared for three workability values ($Sl = 100$ mm, $Sl = 150$ mm and $Sl = 200$ mm) with the same $W/C = 0.35$. The test results of steel fibre-reinforced concrete specimens at three, seven and 28 days are shown in Tables 7.13 and 7.14.

During preparing the mixtures, a clear tendency was observed – it was more difficult to achieve a uniform fibre distribution in concrete with a decrease in the mixture workability. In particular, at $Sl = 100$ mm, it was difficult to achieve the fibre-reinforced concrete mixture homogeneity as fibre clumps formed and, accordingly, significant areas of unreinforced concrete remained. As can be seen from the results (Fig. 7.23), this led to a significant decrease in flexural strength, the value of which was similar to that of unreinforced concrete specimens. The compressive strength does not decrease significantly, but this decrease can also be explained by the concrete structure heterogeneity.

Comparing two types of fibre-reinforced concrete (Figs. 7.23 and 7.24), it should be noted that the fine-grained concrete mixture was less prone to fibre clumping at low workability values. The small size of the aggregate grains

Table 7.13: Strength of fibre-reinforced concrete for the first method of fibre addition

Sl, mm	Compressive strength, MPa			Flexural strength, MPa		
	Three days	Seven days	28 days	Three days	Seven days	28 days
Normal-weight Fibre-reinforced Concrete						
100	55.6	73.3	92.3	1.8	2.6	2.8
150	55.9	72.6	91.4	3.9	5.6	7.8
200	62.3	77.8	96.2	4.8	7.5	8.9
Fine-grained Fibre-reinforced Concrete						
100	53.1	69.8	87.2	2.5	3.4	4.2
150	54.8	68.3	86.8	6.2	8.7	11.3
200	56.2	71.3	93.5	7.5	10.6	13.6

Table 7.14: Strength of fibre-reinforced concrete for the second method of fibre addition

Sl, mm	Compressive strength, MPa			Flexural strength, MPa		
	Three days	Seven days	28 days	Three days	Seven days	28 days
Normal-weight Fibre-reinforced Concrete						
100	56.7	72.4	91.6	1.6	2.1	2.6
150	55.2	73.9	92.4	4	5.3	7.5
200	58	75	93.8	7.6	6.3	7.9
Fine-grained Fibre-reinforced Concrete						
100	53.4	69	86.4	2.6	3.4	4.3
150	54.9	72.1	83.3	6.36	8.5	10.6
200	55.2	70.3	89	7.6	10.8	12.9

allowed more easily fibre distribution in the concrete mixture, which led to an increase in its homogeneity and, thereby, strength indicators.

A comparison of two fibre-reinforced concrete mixture preparation methods shows that both classical and dry methods yield comparable results with a slight predominance of the classical method (mainly for mixtures with high workability), which can be explained by longer overall mixing duration of the mixture components with this method compared to dry. When the workability increased to 150 mm, fibre clumping became less noticeable and at 200 mm, it was not observed at all.

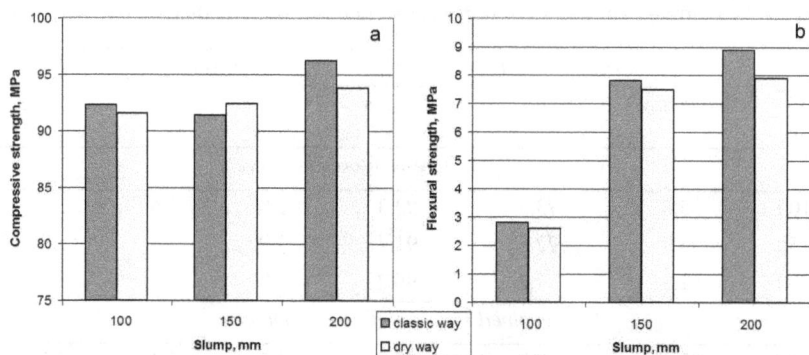

Fig. 7.23: Change in compressive strength (a) and flexural strength (b) of steel fibre-reinforced concrete at 28 days depending on the method of fibre addition and mixture workability

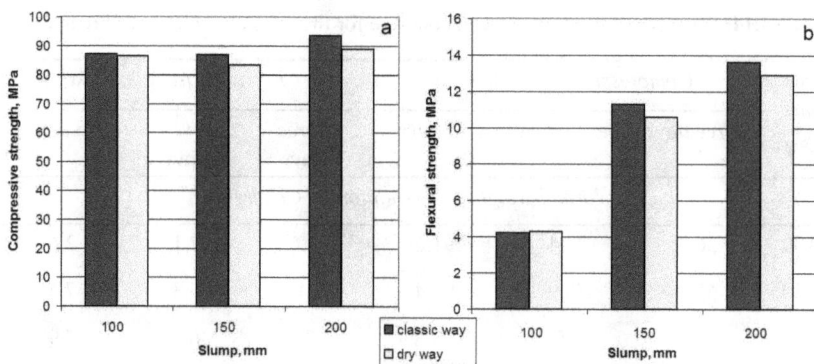

Fig. 7.24: Change in compressive strength (a) and flexural strength (b) of fine-grained steel fibre-reinforced concrete at 28 days depending on the method of fibre addition and mixture workability

Improvement of Fibre-reinforced Concrete Properties by Fibre Orientation

Creating an oriented fibre structure in steel fibre-reinforced concrete, in which the fibre is directed along the tensile stresses acting in the element, is an important technical and scientific task. This orientation yields a significant increase in tensile strength. By now, there are a number of methods to create an oriented fibre structure, most of which can be reduced to the following: fibre compressive by formwork, roller pressing, separate layer-by-layer casting of fibre and concrete mixture, mechanical fibre orientation by passing the mixture through special meshes, fibre spraying into the mixture at high speed, intensive vibration of the mixture, in which the fibre takes a horizontal position, casting concrete elements under a magnetic field.

It was found that fibre orientation by formwork compressive is possible when the concrete elements thickness is less than the fibre length [57]. Calculations show that when the ratio of fibre length to the element thickness is higher than 2, fibre orientation in almost one direction is achieved.

Basic research on roller-pressed steel fibre-reinforced concrete properties was carried out [58]. This type of concrete was suggested to be used for thin-walled elements. In particular, decorative floor tiles with a thickness of 15 cms were produced and tested. It was found that they are characterised by two times lower abrasion compared to ordinary concrete. A rotary-impact forming technology was developed [59]. Following this technology, the mixture is fed from the hopper to a special rotor, which flattens it into thin slabs. A method for casting and compacting fibre-reinforced concrete using special slides was proposed [60]. However, these methods are not suitable for producing elements with a thickness that exceeds the fibre length and, in addition, it is impossible to use these methods to obtain a uniaxial fibre orientation.

It is known that at steel fibre-reinforced concrete mixture vibration, the fibre should take a horizontal position [61]. However, research showed that at long-term intense vibration, delamination of steel fibre-reinforced concrete mixture occurs; as a result, the fibres move into the lower element part [62]. Thus, this orientation method should be used with caution. Another method includes use of spraying of a fibre-reinforced concrete mixture for producing thin-walled elements (less than 2 cms). This yield directs distribution of fibres and for elements with thickness of 3-5 cms, to use conventional casting methods. A method of spraying fibre at high speed at an angle to the concrete mixture was patented [63]. However, such methods are associated with the use of unique expensive equipment and require modernisation of the entire technological process.

Quite exotic methods for producing oriented steel fibre-reinforced concrete elements are also offered; for example, extruding the mixture to ensure a certain fibre orientation, or to pass a special mesh through the mixture [54].

Along with the above-mentioned methods, there is a method of steel fibres' orientation using a magnetic field, based on the property of the fibre to rotate along its force lines [64]. For the first time, the idea of steel fibres' orientation by a magnetic field appeared in the early 70s of the previous century. Steel fibre-reinforced concrete tiles with fibre oriented by permanent magnetic field were produced [64]. Short (1-2 cm) pieces of steel shavings with a thickness of 1-2 mm were used as fibres. Specimens with a flexural strength up to 24 MPa were obtained.

A method of casting steel fibre-reinforced concrete elements, in which the fibres are oriented along the force lines of the field, was proposed [65]. To reduce the field induction necessary for fibres orientation, the fibre-reinforced concrete mixture was simultaneously vibrated on vibrating platform.

The momentum of force required for fibre orientation in the mixture is calculated as follows:

$$M = \tau_0 \, d \, l^2/6 \tag{7.15}$$

where d is the fibre diameter, l is the fibre length, τ_0 is the ultimate shear stress of the mixture.

The power absorbed by the steel fibre-reinforced concrete mixture at vibration under the magnetic field depends on the field cyclic characteristic (ω), reinforcement coefficient (μ), average density of the mixture (ρ) and its viscosity (η) and saturated magnetisation of steel (J):

$$W = \frac{\rho\left(BJd^2\right)\mu\omega}{2\pi(l\eta)} \tag{7.16}$$

The ultimate shear stress of the mixture (τ_0) and mixture viscosity (η) decrease when its workability increases [65]. We have found that it is possible to improve the physical and mechanical properties, namely the fine-grained steel fibre-reinforced concrete flexural strength by the effect of fibre orientation at compaction of concrete mixtures with different workability on a vibrating platform, with and without a magnetic field. The fibre content was 60, 80 and 100 kg/m^3 and the concrete mixture workability varied from 70 to 210 mm. The necessary workability of the concrete mixture was provided by selecting the polycarboxylate type Melflux 2651f superplasticizer content. The results of the research are shown in Figs. 7.25-7.27.

Perpendicular steel fibre orientation to the load direction has a positive effect on physical and mechanical characteristics of fibre-reinforced concrete, which is achieved by vibration compaction of concrete mixtures with high workability and is enhanced by magnetic field (Fig. 7.28).

The obtained results showed that the steel fibre content has the most significant effect on fine-grained steel fibre-reinforced concrete flexural strength: When the fibre content increases from 60 to 100 kg/m^3, the strength increases almost two

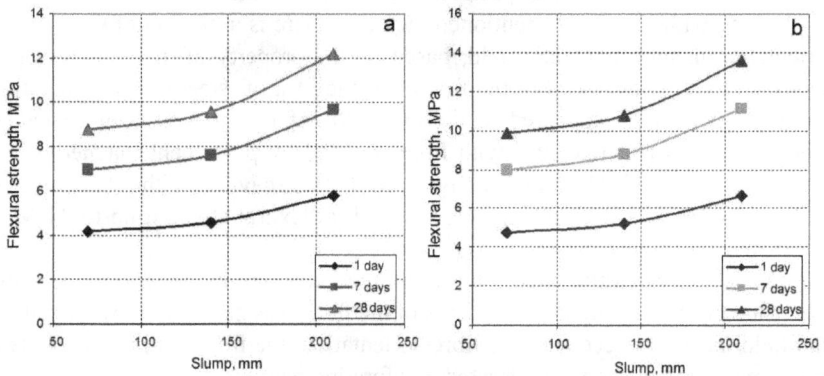

Fig. 7.25: Dependencies of fine-grained steel fibre-reinforced concrete flexural strength on concrete mix workability (Sl) for fibre content of 60 kg/m^3: (a) vibrated without magnetic field; (b) vibrated under magnetic field

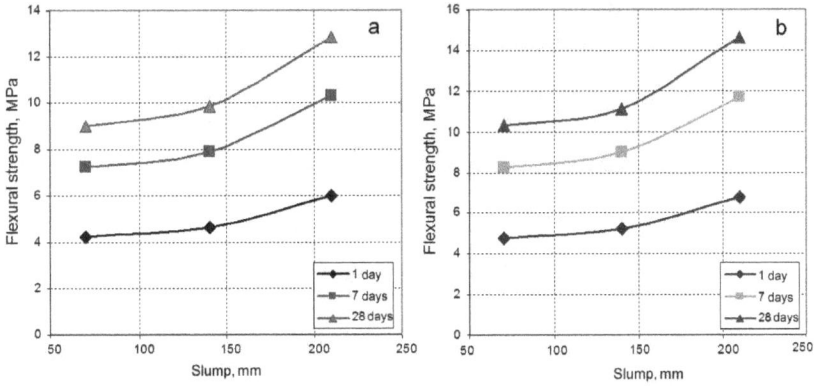

Fig. 7.26: Dependencies of fine-grained steel fibre-reinforced concrete flexural strength on concrete mix workability (*Sl*) for fibre content of 80 kg/m³: (a) vibrated without magnetic field; (b) vibrated under magnetic field

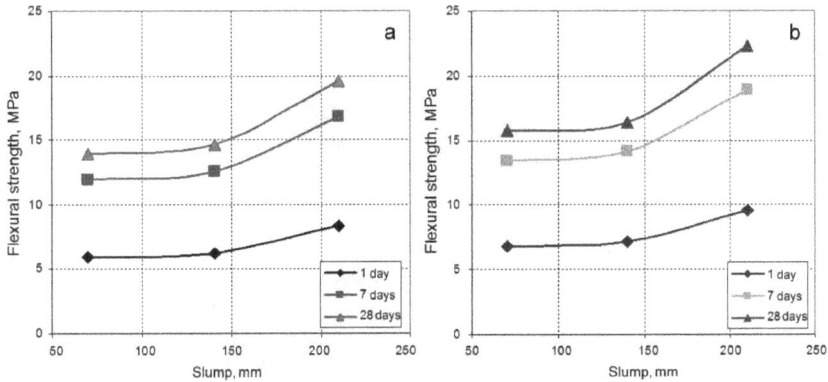

Fig. 7.27: Dependencies of fine-grained steel fibre-reinforced concrete flexural strength on concrete mix workability (*Sl*) for fibre content of 100 kg/m³: (a) vibrated without magnetic field; (b) vibrated under magnetic field

times. If all other conditions are equal, the fine-grained steel fibre-reinforced concrete strength is positively affected by the increase in the concrete mixture workability, which is associated with steel fibre orientation effect strengthening at vibration compaction of concrete mixtures with high workability.

As a result of the increase in workability from 70 to 210 mm, the concrete mixture viscosity decreases and the energy costs required to ensure the steel fibre orientation in one direction are significantly reduced. Experimental studies have demonstrated that if all other conditions are equal, an increase in concrete mixture workability leads to an increase in fine-grained steel fibre-reinforced concrete flexural strength by an average of 40%, which is associated with an increase in the amount of fibre located perpendicular to the applied load. Casting products by vibration compaction under magnetic field leads to some strengthening of

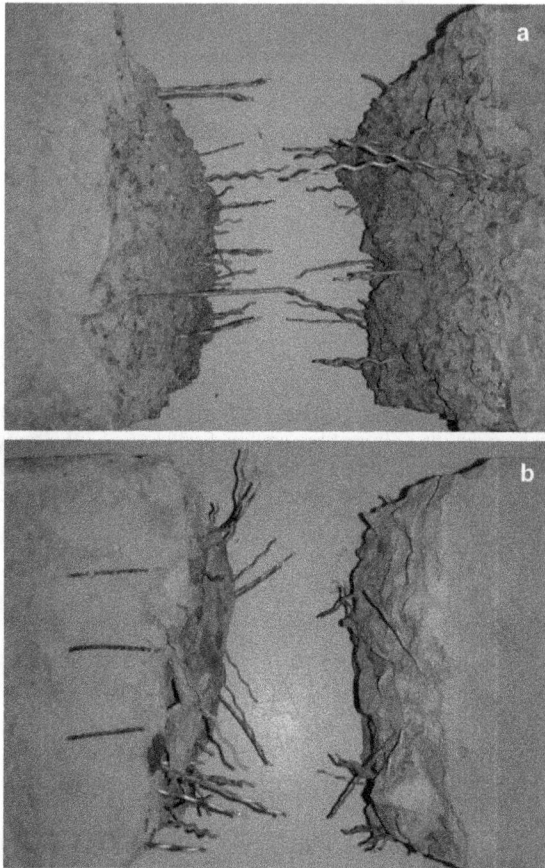

Fig. 7.28: Cross-section view of destroyed specimen with oriented (a) and non-oriented (b) fibre

the steel fibre orientation effect. At the same time, the increase in strength is an average 10-14%, but the concrete mixture workability remains the key factor.

The Ratio of Normal-weight and Fine-grained Fibre-reinforced Concrete Strength Parameters

Results of determining modulus of elasticity for fibre-reinforced concrete [66-78] show that the values of steel fibre-reinforced concrete modulus of elasticity is in the range from 44300 MPa [66] to 110000 MPa [78]. This range can be divided into two groups:

- 44300-57000 MPa for concrete strength classes up to C60;
- 57000-110000 MPa for concrete of strength classes from C80 to C350.

The modulus of elasticity increases:

- with the concrete class growth (from C50 to 80 MPa – from 360 to 600 MPa, from C80 to C145 MPa – from 600 to 11000 MPa);
- when changing the fibre-reinforcement content from 0.5 to 1.5%
- by 0.4-0.6%.

The calculated values of steel fibre-reinforced concrete modulus of elasticity E_{fc} can be obtained based on the precondition that they are proportional to volumetric contents of the concrete matrix and steel fibre having corresponding modulus of elasticity values (E_c and E_f) [75]:

$$E_{fc} = E_c(1 - \mu_{fv}10^{-3}) + E_f\mu_{fv}10^{-3}$$

(7.17)

where μ_{fv} – volumetric content of steel fibre.

The concrete modulus of elasticity is closely related to its deformability and crack resistance. Various criteria have been proposed to characterise the concrete deformability [24]. One of the simplest is the ratio of flexural strength to the statical E_c or dynamic modulus of elasticity (E_d). The ratio between E_c and E_d values for concrete is between 0.87 and 0.95. Lower values of E_c/E_d are typical for concrete with compressive strength less than 25 MPa.

Taking into account the high correlation level of concrete modulus of elasticity with compressive strength, the ratio of concrete tensile strength to compressive one ($f_{c,tf}/f_{cm}$) can be used directly to assess concrete deformability and crack resistance. The value of $f_{c,tf}/f_{cm}$ was obtained for concrete compositions given in Table 7.15 at one, seven and 28 days (Table 7.16).

Table 7.15: Concrete mixtures compositions

No.	W/C	Consumptions of main components, kg/m³				Fibre, kg/m³	Plasticizer type and content, %
		W	C	Sand	Crushed stone		
Normal-weight Fibre-reinforced Concrete							
1	0.46	230	500	655	1252	–	–
2	0.48	238	500	655	1252	80	–
3	0.36	178	500	655	1252	80	C-3 (1%)
4	0.27	135	500	655	1252	80	Melflux (0.5%)
Fine-grained Fibre-reinforced Concrete							
5	0.53	263	500	837	1023	–	–
6	0.50	248	500	837	1023	100	–
7	0.38	188	500	837	1023	100	C-3 (1%)
8	0.32	161	500	837	1023	100	Melflux (0.5%)

This ratio, as is known [78], increases as the concrete structure homogeneity increases and the number of various defects contributing to stress concentration decrease. The ratio increases also significantly for concrete with dispersed reinforcement. All the factors contributing to the cement stone bonding with aggregates and fibre have the highest influence on the fibre-reinforced concrete tensile strength. Graphical dependencies of the ratio between flexural and compressive strength are shown in Fig. 7.29.

Analysing the experimental results given in Table 7.16 and Fig. 7.29, it can be noted that along with a general tendency to decrease $f_{c,tf}/f_{cm}$ over time, for fine-grained fibre-reinforced concrete this tendency is much more pronounced with addition of superplasticizer . This is explained by a more intensive growth of compressive strength over time as compared to flexural strength.

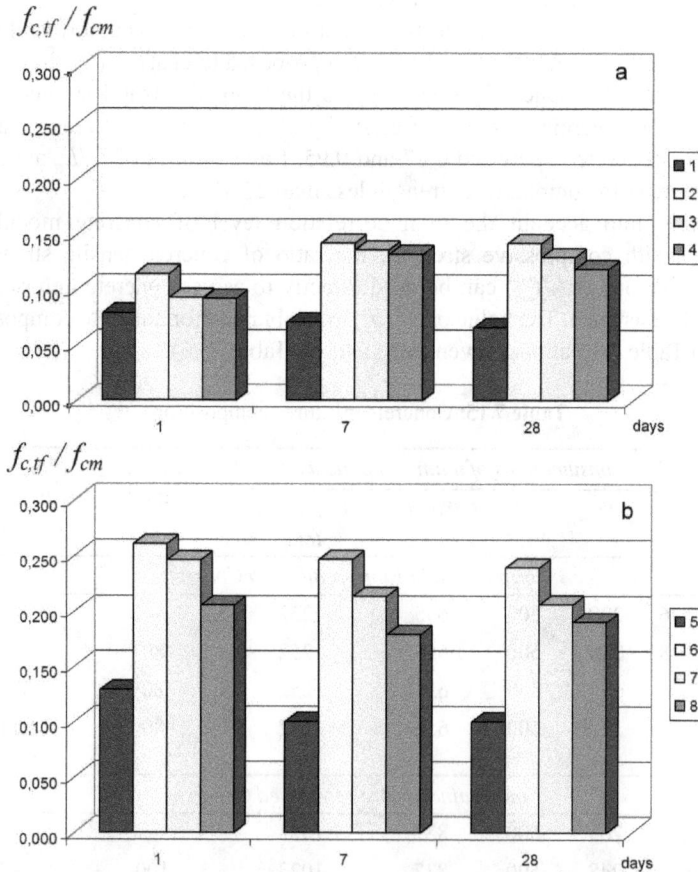

Fig. 7.29: Ratio of flexural to compressive strength of normal-weight (a) and fine-grained (b) fibre-reinforced concrete (column numbers correspond to the compositions given in Table 7.15)

Table 7.16: Values of $f_{c,tf}/f_{cm}$ ratio for normal-weight and fine-grained concrete and fibre-reinforced concrete with optimal fibre content

Concrete composition number	Concrete strength indicators, MPa and their ratio at								
	One day			Seven days			28 days		
	$f_{c,tf}$	f_{cm}	$f_{c,tf}/f_{cm}$	$f_{c,tf}$	f_{cm}	$f_{c,tf}/f_{cm}$	$f_{c,tf}$	f_{cm}	$f_{c,tf}/f_{cm}$
Normal-weight Concrete									
1	1.5	18.6	0.079	2.0	28.7	0.071	3.5	53.2	0.066
2	2.6	22.8	0.114	4.7	32.8	0.142	7.9	55.6	0.142
3	3.6	38.4	0.093	8.9	65.1	0.136	10.8	83.5	0.129
4	4.2	45.9	0.092	10.8	81.7	0.132	12.1	102.1	0.119
Fine-grained Concrete									
5	2.1	16.1	0.129	3.2	31.7	0.100	4.6	45.9	0.100
6	5.7	21.8	0.260	7.8	31.5	0.247	11.6	48.5	0.239
7	7.9	32.0	0.246	12.2	57.0	0.213	14.3	69.5	0.206
8	8.6	41.8	0.205	12.6	70.8	0.178	16.2	85.3	0.190

So, if for fine-grained fibre-reinforced concrete without plasticizers' compressive strength at 28 days is 48.5 MPa, the $f_{c,tf}/f_{cm}$ ratio at one day is 0.26; at seven days, it is 0.247; at 28 days, it is 0.239, then for fibre-reinforced concrete with C-3 superplasticizer , the values are 0.246, 0.213 and 0.206, respectively. For fine-grained fibre-reinforced concrete with Melflux, these ratios are 0.205, 0.178 and 0.190, respectively.

Technological Properties of Fibre-reinforced Concrete Mixtures

Water demand and workability are the most important interrelated technological properties of concrete mixtures, which determine both their ability to compact and, to a large extent, the hardened concrete properties. The water demand constancy rule is well known in concrete technology [31]. In accordance with this rule, at constant water content, the cement consumption in the range between 200 and 400 kg/m^3 has no significant effect on concrete mixture workability. The concrete mixture's water demand required for achieving a given workability index is practically constant in certain cement consumption and W/C ranges.

Finding the upper limit of the water demand constancy rule region, which allows taking into account the used cement peculiarities, is achieved at the critical W/C ($(W/C)_{cr}$), which is equal on an average to 1.68 $K_{n.c}$, where $K_{n.c}$ is the W/C, which corresponds to the cement paste normal consistency [34]. For normal-weight concrete, $(W/C)_{cr}$ is within 2.2-2.4.

From a physical viewpoint, the water demand constancy rule is that with an increase in C/W to a certain critical value, the increase in the cement paste structural viscosity in the concrete mixture is compensated by an increase in its quantity and, accordingly, the cement paste layer thickens on the aggregate grains. Beyond the critical C/W, an increase in the amount of cement 'lubricant' no longer compensates for the progressively increasing water demand of the concrete mix.

To study the effect of C/W on water demand of high-strength fibre-reinforced concrete, experiments were carried out at the optimal F1 fibre content, which was determined before. For normal-weight fibre-reinforced concrete, the fibre content was 80 kg/m^3 ($\mu = 1\%$), for fine-grained, it was 100 kg/m^3 ($\mu = 1.3\%$). Concrete mixtures were prepared at four different values of C/W. The mixtures' workability was maintained within 150 mm. The experimental results are shown in Table 8.1 and Fig. 8.1.

Table 8.1: Water content and mixture workability depending on cement-water ratio

C/W	W/C	Water content, l	Slump (Sl), mm
Normal-weight Fibre-reinforced Concrete			
1.8	0.55	205	140
2.2	0.45	206	160
2.7	0.37	231	150
3.1	0.32	256	150
Fine-grained Fibre-reinforced Concrete			
1.8	0.55	220	150
2.2	0.45	223	160
2.7	0.37	245	140
3.1	0.32	285	150

Following the results, the water demand constancy rule is also valid for using fibre-reinforced concrete mixtures. The upper limit of the 'critical' C/W region for such mixtures is within 2.2-2.3 (Fig. 8.1). Considering similar mixtures' compositions without fibre, it can be noted that adding fibre leads to an increase in water demand by 3-12%, depending on the aggregate type. Fine-grained mixtures are characterised by a higher water demand compared to mixtures on coarse aggregate, which is explained by higher total surface of the grains in the first case.

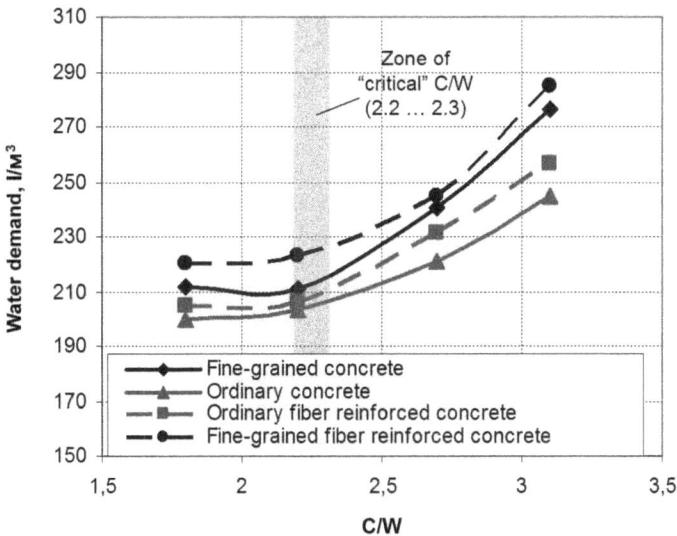

Fig. 8.1: The effect of C/W on water demand of concrete mixtures

It should also be noted that normal-weight fibre-reinforced concrete mixtures with high C/W values were characterised by a tendency to delamination while for fine-grained mixtures, this phenomenon was expressed to a lesser extent.

Considering the influence of fibre content on concrete mixture water demand (Fig. 8.2), it is possible to note constant water demand growth as fibre volumetric content becomes higher.

The use of Melflux and C-3 plasticizing admixtures, along with the reduction of water demand, increases the 'critical' C/W to values 2.6-2.7 (Fig. 8.3).

Within the water demand constancy rule limits for the tested concrete, the dependence of concrete mixture workability on water content was obtained (Fig. 8.4). The consumption of Melflux superplasticizer was 0.5%, C-3 was 1%. Fibre content corresponded to previously obtained optimal values. As can be seen from the above data, the use of superplasticizer leads to the fact that mixtures (especially fine-grained ones) become more sensitive to changes in workability even for small changes in water content.

To study the change in water demand of concrete mixtures outside the water demand constancy rule, increase in water demand ΔW can be found using the following empirical formula [34]:

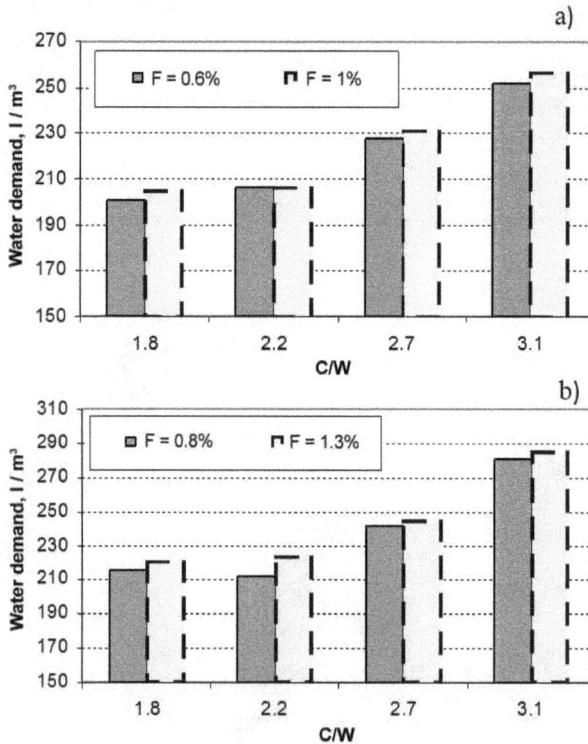

Fig. 8.2: Influence of fibre content and C/W on water demand of mixtures for normal-weight (a) and fine-grained (b) fibre-reinforced concrete

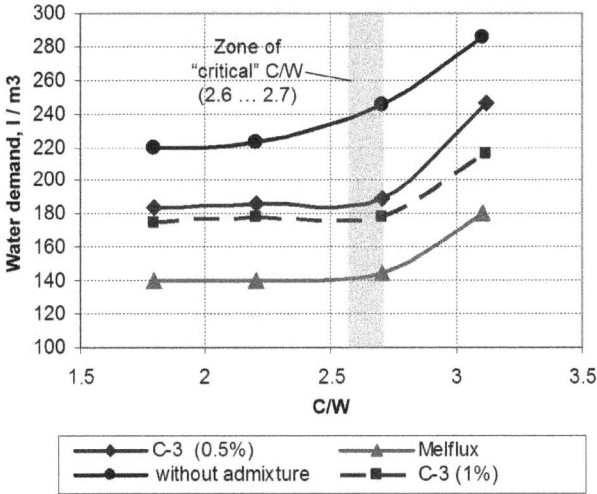

Fig. 8.3: The effect of *C/W* on water demand of fibre-reinforced concrete mixtures when using water-reducing admixtures

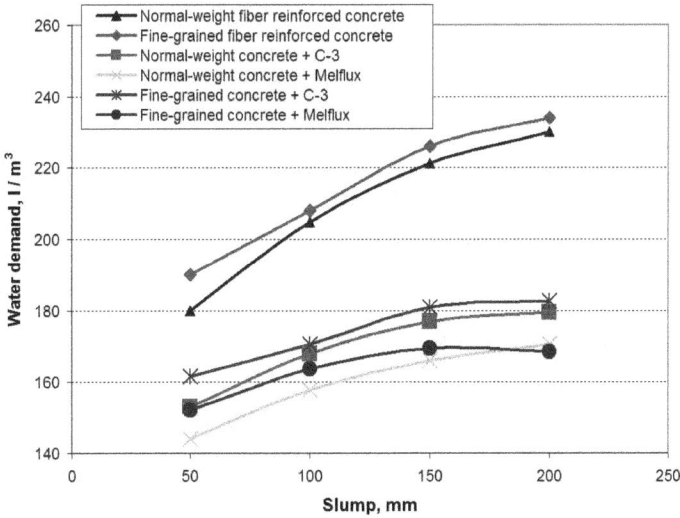

Fig. 8.4: The influence of workability on water demand of concrete mixtures when using superplasticizers

$$\Delta W = \left(W/C - \frac{1}{1.68K_{n.c}} \right) \left(\frac{W_0}{100} \right)^{5.5} \tag{8.1}$$

where W_0 is water demand; obtained the water demand constancy rule limits; $K_{n.c}$ is the normal consistency of cement.

Table 2 shows the values of concrete mixtures' water content when $C/W >$ $(C/W)_{cr}$ and various indicators of concrete mixture workability, as well as the values of ΔW, calculated according to Eq. (8.1) and obtained experimentally. The fibre-reinforced concrete mixtures' water content was found under the condition of using the basic compositions of concretes, which were obtained earlier.

The obtained results show good convergence in terms of correction values, both calculated by Eq. (8.1) and found experimentally. The values of corrections given in Table 8.2 can be used for design of high-strength fibre-reinforced concrete compositions that require high C/W values.

Table 8.2: Calculated and experimental values of the correction to fibre-reinforced concrete mixtures' water demand

Slump, mm	$W_0, l/m^3$ $(C/W)_{cr} \leq 2.2$	Correction water demand $\Delta W,$ l/m^3 (according to Eq. 8.1)		Correction water demand $\Delta W, l/m^3$ (Experimental)	
		$C/W = 2.7$	$C/W = 3.1$	$C/W = 2.7$	$C/W = 3.1$
Normal-weight Fibre-reinforced Concrete					
50	185	14.6	26.4	15	27
100	205	25.7	46.4	24	42
150	218	36.0	65.1	32	61
200	225	42.9	77.5	43	75
Fine-grained Fibre-reinforced Concrete					
50	192	17.9	32.4	16	31
100	210	29.3	53.0	26	50
150	225	42.9	77.5	43	76
200	238	58.4	105.5	56	103

Saving Properties of Fibre-reinforced Concrete Mixtures over Time

An important issue for investigating the technological properties of fibre-reinforced concrete mixtures is prediction of the workability loss – the so-called sustainability, or mixture preservation over time, allowing corrections when assigning the initial workability and determine the permissible transportation duration of the mixture. For concrete and fibre-reinforced concrete mixtures with high slump, as workability preservation indicator can be time taken during which the average cone slump value decreases from 200 to 150 mm. For a comparative assessment of the mixtures workability saving over time, tests of concrete and fibre-reinforced concrete mixtures were carried out for compositions presented in Table 7.15. Figure 8.5 presents kinetics of workability index saving for concrete mixtures compositions with coarse aggregate as specified in Table 7.15. Mixtures with superplasticizer have the highest loss of workability over

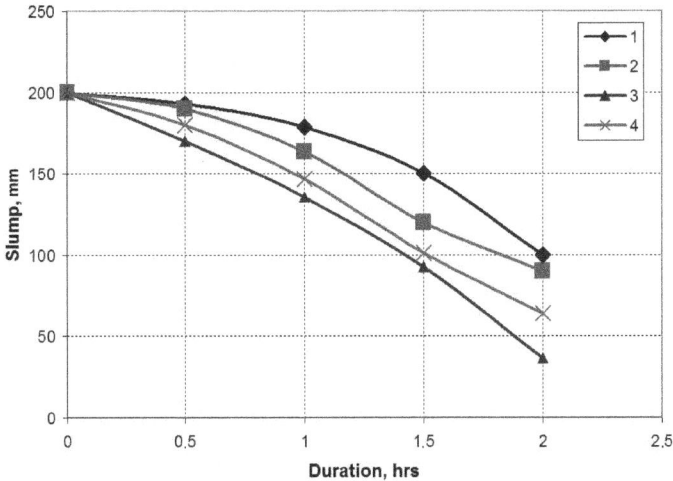

Fig. 8.5: Kinetics of changes in normal-weight concrete mixtures workability
(the compositions numbers correspond to Table 7.15)

time and without plasticizer and use of fibre, it is the lowest. It is known that plasticized flowable mixtures lose their workability faster than uniform mixtures without superplasticizers, which is naturally explained by higher water content of the latter.

Addition of fibre leads to a certain decrease in the mixtures' workability saving over time compared to that without fibre. In general, workability losses during the first hour for all compositions were insignificant. The mixtures' workability saving for compositions 1-4 is 1.5 h, 1.35 h, 0.95 h and 0.76 h, respectively.

Considering the kinetics of changes in workability over time for fine-grained concrete (Fig. 8.6), the following features can be noted. In general, the nature of the dependencies is similar to normal-weight concrete; at the same time, in absolute terms, the workability loss is more significant. The highest loss of workability over time is also experienced for mixtures with superplasticizers and the kinetics of changes in workability of mixture compositions with both used superplasticizers is almost the same up to 1.5 h, the lowest, without superplasticizer and fibre.

Losses of workability during the first hour for all compositions were more significant compared to mixtures on coarse aggregate. Workability stability of mixtures' compositions 5-8 was 1.34 h, 0.77 h, 0.51 h and 0.49 h, respectively.

When studying the process of change workability, a characteristic feature of mixtures with superplasticizers was noticed. The shelf-life (viability) of such mixtures can be significantly increased by using additional constant or periodic mixing after their preparation. Figure 8.7 shows comparison of kinetics in workability changes over time for fine-grained fibre-reinforced concrete mixtures (compositions 7 and 8, Table 7.15) mixed once and the same mixtures that were subjected to additional mixing of 2-2.5 min. every 0.5 h.

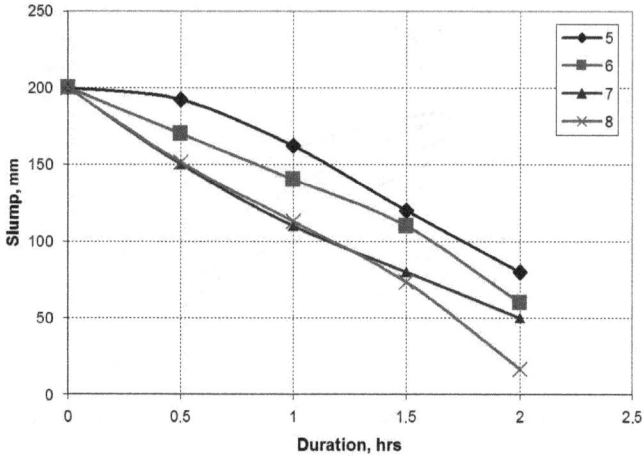

Fig. 8.6: Kinetics of changes in fine-grained concrete mixtures workability
(the compositions numbers correspond to Table 7.15)

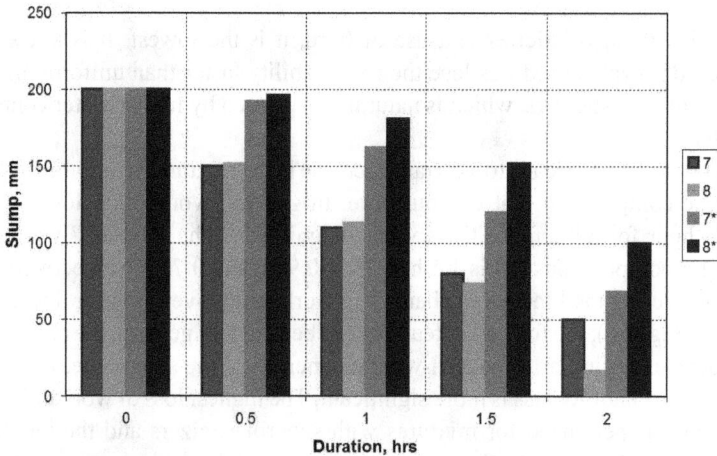

Fig. 8.7: Kinetics of changes in fine-grained concrete mixtures workability with single
and repeated mixing (the compositions numbers correspond to Table 15, * with
additional mixing)

As can be seen from the above data, additional mixing can significantly increase the workability preservation of fibre-reinforced concrete mixtures with the use of plasticizing admixtures. This is especially relevant for mixtures with addition of polycarboxylate superplasticizer Melflux, for which the lifetime (with a decrease in workability by 5 cm) was 1.54 h, which is three times higher than that for a mixture that was mixed just once. To increase the life-time of fibre-reinforced concrete mixtures, it is advisable to add superplasticizers after preliminary (2 minutes) mixing of the concrete mixture. This method enables to obtain some savings in the admixture compared to mixing it with water to obtain mixtures and concrete with the same characteristics.

The duration of plasticizing admixture action also increases when it is added into the concrete mixture in portions. The effectiveness of such addition for obtaining mixtures with high workability can be explained from the viewpoint of the need to maintain a certain excess amount of plasticizer in the hydrated cement-liquid phase. However, repeated introduction of an admixture to restore workability can lead to the pore structure deterioration and, as a result, a decrease in concrete technological properties.

Technological methods related to casting, compaction and processing of concrete reinforced with dispersed fibres are practically similar to traditional ones. When producing dispersed reinforced concrete, increased attention should be paid to vibration duration, which has a significant effect on fibre distribution uniformity in concrete volume. Exceeding the concrete mixture vibration duration beyond the set time (depending on its composition) can lead to delamination of the reinforced mixture. In this case, due to the difference in the density of concrete and steel, under the gravity forces action, the fibre tends to fall down (to the pallet) during the mixture vibration. Sometimes this property is used to provide zonal reinforcement, when a higher reinforcement level is required according to calculation in any part (zone) of the product. Adjusting the vibration duration to ensure uniform or zonal reinforcement is done at the products' manufacturing stage [79].

To avoid delamination of steel fibre-reinforced concrete mixture, experimental studies were performed to investigate the influence of steel fibre content and vibration compacting duration on delamination of concrete mixture with different workability. All studies were performed on fine-grained concrete of the same composition; the necessary concrete mixture workability was provided by selecting the Melflux polycarboxylate type superplasticizer content. The steel fibre content was 60, 80 and 100 kg/m^3 of the concrete mixture.

The fibre distribution uniformity in the concrete mixture volume was estimated by delamination coefficient, which was determined according to the known method [80]. This technique includes placing a fibre concrete mixture in a cylindrical form with a height of 200 mm and a diameter of 100 mm and further vibration compaction. After that, the specimen is cut into two equal parts – upper and lower. From each of the parts, fibre is separated by washing and weighting. The delamination coefficient is calculated as:

$$K_d = m_{top.f} / m_{down.f} \qquad (8.2)$$

where $m_{top.f}$ and $m_{down.f}$ are weights of fibre from the upper and lower parts of the specimen.

Results of experimental studies on fine-grained reinforced concrete coefficient delamination are shown in Table 8.3.

The obtained experimental results indicate that the most significant factors affecting the delamination coefficient of steel fibre-reinforced concrete mixture

Table 8.3: The effect of compaction vibration duration and concrete mixture
workability of delamination coefficient of fine-grained steel fibre-reinforced concrete
with a fibre content of 60-100 kg/m^3

Mixture workability (Sl), mm	*Delamination coefficient for vibration compaction duration of sec.*			
	15	*30*	*45*	*60*
Fibre Content 60 kg/m^3				
50-90	0.96	0.91	0.85	0.76
100-150	0.88	0.82	0.75	0.64
160-210	0.83	0.71	0.58	0.47
Fibre Content 80 kg/m^3				
50-90	0.94	0.9	0.83	0.74
100-150	0.86	0.79	0.72	0.63
160-210	0.81	0.7	0.57	0.46
Fibre Content 100 kg/m^3				
50-90	0.91	0.86	0.8	0.71
100-150	0.84	0.76	0.7	0.6
160-210	0.79	0.65	0.55	0.44

are its workability and compaction vibration duration. It was found that when the concrete mixture workability increases, it is necessary to significantly limit the compaction vibration duration.

To ensure the necessary delamination coefficient, which should be up to 0.8 for mixes of S3 workability (slump) class and up to 0.75 for S4 and S5 workability classes [80], the optimal compaction vibration duration should be up to 45 sec. for mixtures with concrete slump of 50-90 mm, 15-30 sec. for 100-150 mm and 15 sec. for 160-210 mm. It was also demonstrated that increasing the fibre content in the investigated range has no significant effect on delamination coefficient of fine-grained steel fibre-reinforced concrete. At constant concrete mixture workability and compaction vibration duration, an increase in steel fibre content from 60 to 100 kg/m^3 of the concrete mixture leads to delamination coefficient decrease by 3-5%.

Thus, the performed experimental studies enabled to establish optimal compaction vibration duration of fine-grained steel fibre-reinforced concrete, providing uniform distribution of steel fibre in the concrete volume.

Properties of Fibre Concrete Determining Its Durability

Voidness and Water Absorption

Low water content of the investigated fibre-reinforced concrete mixtures, intensive hydration and hardening determine the corresponding features of the concrete pore structure. When studying the qualitative features of concrete pore structure, a method based on water saturation kinetics analysis is widely used. This method is based on empirically established relationship between parameters of exponential function that characterises concrete water absorption over time with parameters of concrete porosity.

The curves characterising the change in water absorption of normal-weight concrete vs. hardening duration are satisfactorily approximated by functions having the following general form [81]:

$$W_\tau = W_{max}\left[1 - e^{-(\lambda\tau)^\alpha}\right] \tag{9.1}$$

where W_τ is water absorption of the specimen over time τ; W_{max} is maximum water absorption; λ is coefficient characterising the average capillaries size; α is coefficient of capillary sizes uniformity.

To calculate the coefficients λ and α depending on the values of W_τ, W_{max} and τ corresponding nomograms are proposed [82].

The results of experiments and calculations are given in Table 9.1 and in Fig. 9.1.

Analysis of the data given in Table 9.1 and in Fig. 9.1 shows that, as expected, decreasing the concrete mixtures' water content and water-cement ratio resulted in a significant decrease in water absorption of fibre concrete as a characteristic of its open capillary porosity.

Theoretically, this conclusion follows from analysis of equation for concrete capillary porosity:

$$P_{cap} = \frac{W - 0.5\alpha C}{1000} \qquad (9.2)$$

where α is cement hydration degree; C is cement consumption and W is water content, kg/m^3.

Table 9.1: Parameters of concrete pore structure

Composition number	W/C	Water content kg/m³	Total voidness, % (V)	Open capillary voidness, % (W₀), % (W₀)	Average voids size index λ	Homogeneity indicator of voids, α	Fibre, kg/m³	Superplasticizer type and content, %
Normal-weight Concrete								
1	0.46	230	11.6	7.5	1.61	0.78	–	–
2	0.48	238	11.8	7.4	1.55	0.81	80	–
3	0.36	178	9.7	5.1	1.46	0.85	80	C-3 (1%)
4	0.27	135	9.9	4.9	1.41	0.88	80	Melflux (0.5%)
Fine-grained Concrete								
5	0.53	263	12.4	8.5	1.65	0.75	–	–
6	0.50	248	12.1	8.1	1.53	0.82	100	–
7	0.38	188	9.3	5.5	1.43	0.87	100	C-3 (1%)
8	0.32	161	9.5	5.3	1.39	0.89	100	Melflux (0.5%)

Notes: 1. Numbers of concrete compositions correspond to Table 7.15.
2. Water temperature at concrete testing for water absorption $t = 20$ °C.

There is also a clear correlation between the above-mentioned characteristics and pore structure parameters λ and α. A decrease in the average pore size index is accompanied by increase in structure orderliness, as evident from increase in homogeneity index α.

Kinetics of water absorption growth for all types of concrete can be described by logarithmic relationships. Fine-grained fibre concrete is characterised by lower indicators of average pore size λ at practically equal homogeneity indicators values (Table 9.1).

An important consequence of adding superplasticizer into fibre-reinforced concrete mixtures is improvement of the pore space structure by reducing the

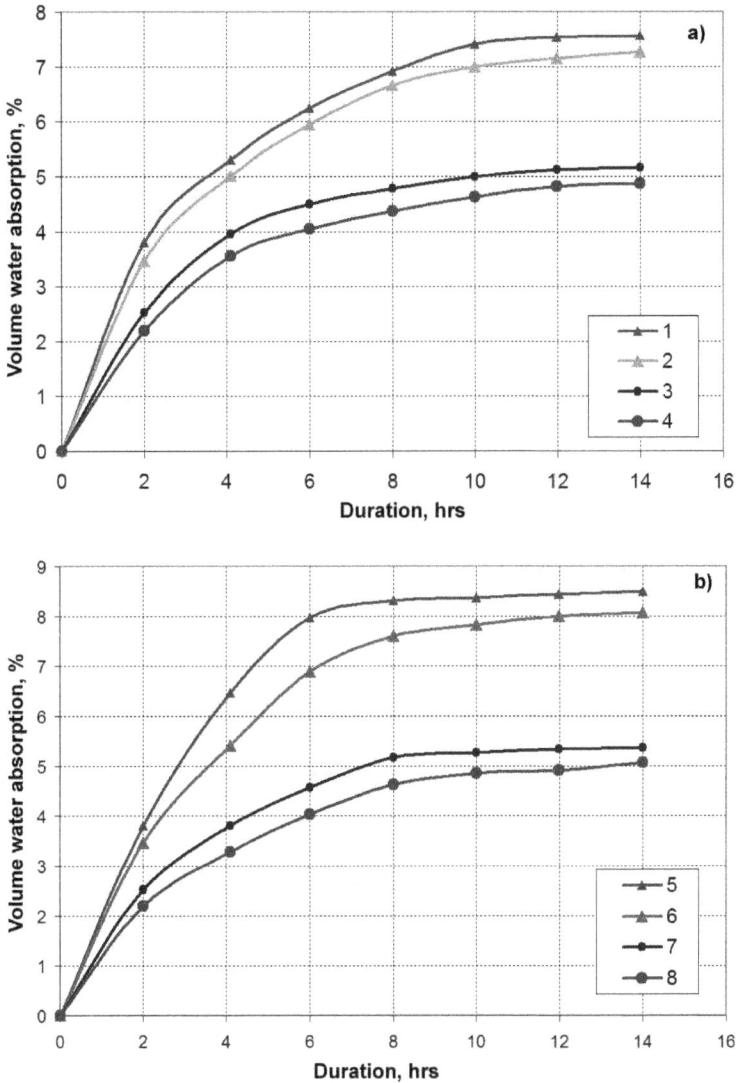

Fig. 9.1: Kinetics of changes in water absorption for normal-weight (a) and fine-grained (b) concretes (the curves numbers correspond to compositions, given in Table 7.15)

average size of pores and increasing the homogeneity of their distribution. A decrease in the water layers' thickness on the binder grains causes a decrease in average radius of capillaries [24]. The reduction in pore radius and their distribution homogeneity is also positively affected by high specific surface area of the binder. Adsorption on active centres of the solid phase contribute to reduction in pore size of concrete with superplasticizers and their distribution homogeneity.

Shrinkage Deformations

Crack resistance of concrete and fibre-reinforced concrete is largely determined by shrinkage deformations (ε_{sh}).

Concrete shrinkage is affected by numerous factors, the dominant one is water demand. At constant water demand in the mixture, the value of ε_{sh} is slightly dependent on cement consumption and C/W [9]. At the same time, the positive role of concrete structure dispersed reinforcement on shrinkage deformations reduction is known.

Shrinkage deformations of concrete and fibre-reinforced concrete compositions given in Table 9.2 were measured. The specimens were stored at a temperature of $20 \pm 2°C$ and a relative humidity of $75 \pm 5\%$. The shrinkage deformation curves are shown in Fig. 9.2. Analysis shows that for all studied compositions, shrinkage deformations stabilise at 60 days.

Ultimate shrinkage deformations of concrete, reached up to 60 days, depending on the composition range from 5.2 to 7.8 ($\times 10^{-4}$). The lowest slump values are typical for normal-weight fibre-reinforced concrete with aggregate size of up to 20 mm, minimum water demand (135 l/m^3), and superplasticizer. At a constant water demand, shrinkage deformations slightly increase with an increase in the superplasticizer content. At seven days, shrinkage reaches 30-40% of the ultimate value, at 28 days, it is 50%, at 60-90 days, it is 85-95%.

In general, shrinkage deformations for fine-grained concrete are 10-15% higher than for normal-weight concrete, which is explained by higher water demand of the first. Adding fibre to optimal content from the strength characteristics

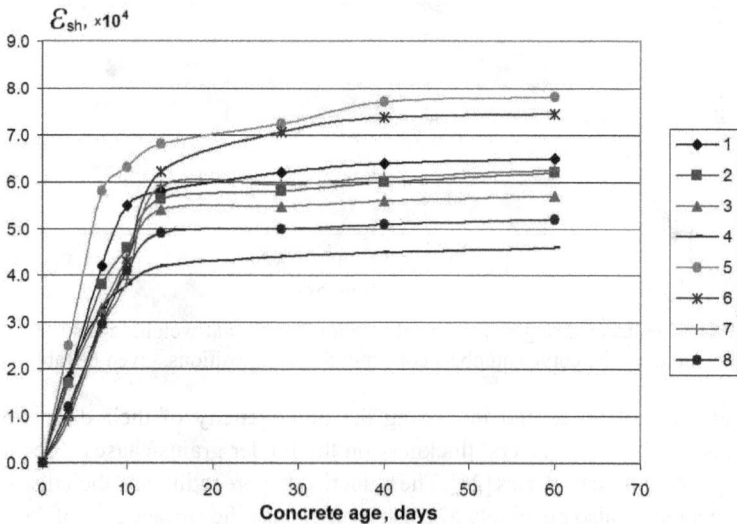

Fig. 9.2: Shrinkage deformations curves (curves numbers correspond to compositions given in Table 7.15)

Table 9.2: Concrete shrinkage deformations and crack resistance modulus

Composition number	Values of ε_{sh} and modulus T for concrete at					
	28 days			90 days		
	$\varepsilon_{sh}\cdot 10^4$	$f_{c,tf}{}^*$, MPa	$f_{c,tf}/\varepsilon_{sh}{}^{**}$	$\varepsilon_{sh}\cdot 10^4$	$f_{c,tf}{}^*$, MPa	$f_{c,tf}/\varepsilon_{sh}{}^{**}$
1	6.2	3.5	0.56	6.5	3.6	0.55
2	5.8	7.9	1.36	6.2	7.9	1.27
3	5.5	10.8	1.97	5.7	11	1.93
4	4.4	12.1	2.75	4.6	12.4	2.70
5	7.2	4.6	0.64	7.8	4.4	0.56
6	7.06	11.6	1.64	7.45	11.7	1.57
7	5.95	14.3	2.40	6.25	14.5	2.32
8	5.00	16.2	3.24	5.2	16.3	3.13

* $f_{c,tf}$ – Flexural tensile strength of concrete

** $f_{c,tf}/\varepsilon_{sh}$ – Modulus crack resistance ($M_{c.r.}$)

viewpoint has a positive effect on shrinkage reduction, although this effect is not significant (curves 1, 2, 5 and 6).

To study the influence of F1 fibre content on shrinkage deformations, tests were carried out for fine-grained fibre-reinforced concrete with fibre content varying from 0 to 3% (Fig. 9.3). The influence of fibre type F2 on fine-grained fibre concrete shrinkage is demonstrated in Fig. 9.4. Depending on fibre type and reinforcement ratio, steel fibre-reinforced concrete shrinkage deformations can be equal, more or less than those of the concrete matrix (Figs. 9.3 and 9.4). It can be explained by the influence of two different processes on reinforced concrete shrinkage. One of them creates conditions for increasing the steel fibre-reinforced concrete shrinkage compared to the matrix and the other prevents shrinkage deformations development. At the same time, the influence degree of these processes depends on fibre type and content.

Anchor shape of fibre type F2 loosens the steel fibre-reinforced concrete mixture at mixing and casting, creating pores filled by steam-air mixture, further evaporation of which leads to increase in shrinkage. Corrugated fibre of type F1, which has good adhesion along its entire length with the hardened matrix, reduces shrinkage deformations of the matrix. Shrinkage deformations also decrease at higher reinforcement ratio.

Figure 9.5 shows the values of shrinkage deformations and crack resistance modulus at 28 days for the investigated concrete and fibre-reinforced concrete compositions. It is evident that all compositions with minimum value of ε_{sh} have the maximum value of modulus $M_{c.r.}$.

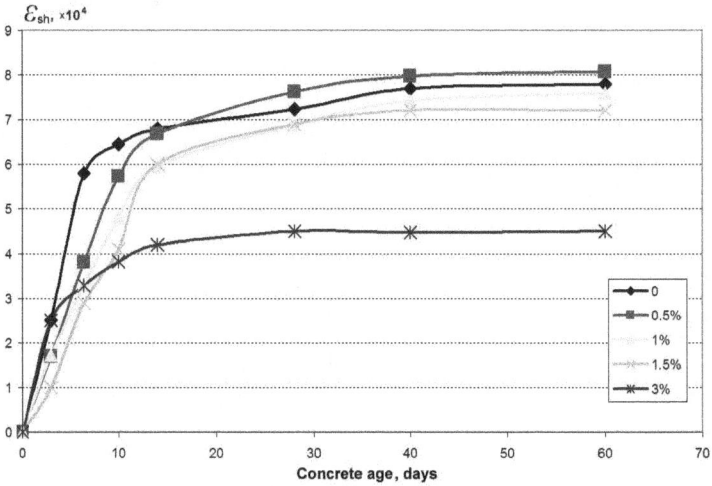

Fig. 9.3: Shrinkage deformations curves of fine-grained fibre concrete for fibre content from 0 to 3%

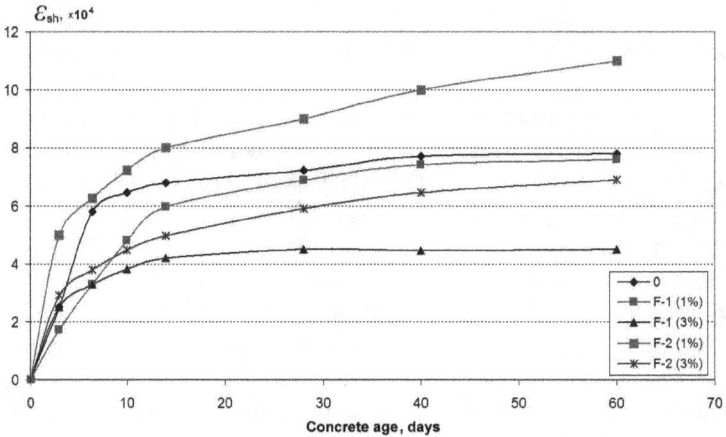

Fig. 9.4: Shrinkage deformations curves for fine-grained fibre-reinforced concrete vs. fibre content for fibre types F1 and F2

Frost Resistance

According to known concrete science representations [9], concrete frost resistance is determined by a complex of factors and first of all, by their capillary porosity as well as the open capillary and closed air pores volume ratio. Along with cement clinker, chemical and mineralogical composition have a significant impact on mineral admixtures type and content. At the same time, for fibre-reinforced concrete, the nature and type of dispersed reinforcement is of high importance [79]. Fibres, increasing concrete tensile strength, provide better counteraction to the stresses that arise during water freezing in capillaries.

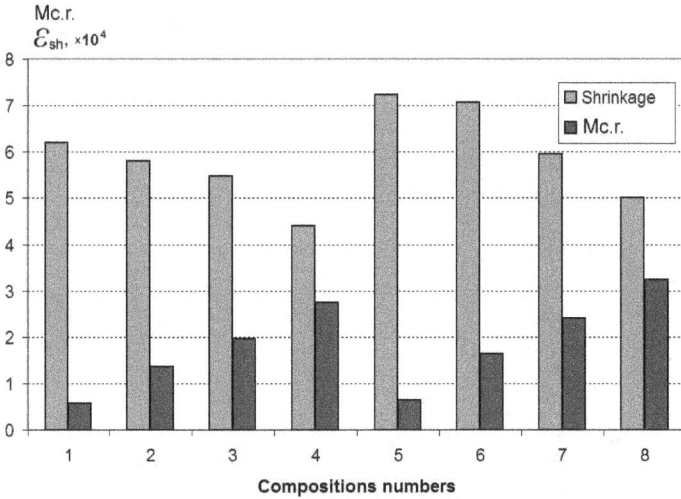

Fig. 9.5: Shrinkage deformations (ε_{sh}) and modulus of crack resistance $M_{c.r} = f_{c,tf}/\varepsilon_{sh}$ at 28 days for the investigated concrete compositions (compositions numbers correspond to Table 7.15)

It is known [36] that the use of composite mineral binders enables to bring the capillary porosity in concrete mixtures with a polycarboxylate superplasticizer to 5%, which, according to known recommendations, should provide sufficiently high frost resistance. At the same time, data on the influence of superplasticizer on concrete frost resistance are contradictory. It was reported about high frost resistance of concrete with addition of C-3 naphtalene-formaldehyde superplasticizer at low W/C as a result of cement stone and concrete structure compaction and a decrease in capillary porosity in proportion to that in mixing water content [36]. However, according to other researchers, in concrete with superplasticizers the pore structure deteriorates and the size of capillaries and the distance between closed pores increase. According to Japanese researchers [78], satisfactory frost resistance of high-strength concrete with $W/C = 0.3$ is obtained with air entrainment of at least 5%. These considerations served as the basis for recommendations [36] to add air-entraining admixtures in concrete with high frost resistance along with the superplasticizer.

Our experiments (Table 9.1) have shown that, along with the decrease in capillary porosity, a decrease in their size and in the pores distribution homogeneity are characteristic for the investigated fine-grained steel fibre-reinforced concrete.

Steel fibre-reinforced concrete is characterised by sufficiently high frost resistance values, which are affected both by characteristics of concrete pore structure, nature and degree of dispersed reinforcement. The frost resistance for such concrete usually ranges from F100 to F600, and for high-strength concrete it can reach F1000 and more [83].

Table 9.3: Air entrainment of concrete mixtures

W/C	Fibre, kg/m³	Superplasticizer type and content, %	Entrained air content, ($V_{e.a.}$), %	Criterion $SD = \dfrac{V_{cap.por}}{V_{cap.por} + V_{e.a.}}$
Normal-weight Concrete				
0.46	–	–	1.4	0.84
0.48	80	–	2.7	0.73
0.36	80	C-3 (1%)	2.2	0.70
0.27	80	Melflux (0.5%)	3.1	0.61
Fine-grained Concrete				
0.53	–	–	1.3	0.87
0.50	100	–	2.4	0.77
0.38	100	C-3 (1%)	2.1	0.72
0.32	100	Melflux (0.5%)	2.9	0.65

For the studied concrete mixture compositions, air entrainment was determined by the compression method. Knowing the open capillary pores volume, which is equal to the volumetric water absorption ($V_{cap.por} = W_0$) and the freezing water volume ($V_{fr.w}$) as well as the entrained air volume ($V_{e.a.}$), the saturation degree (SD) was calculated to be similar to the first structural criterion of frost resistance proposed by T. Whiteside and H. Sweet [84]:

$$SD = \frac{V_{fr.w}}{V_{fr.w} + V_{e.a.}} \qquad (9.3)$$

It was found that at SD < 0.88 concrete is frost resistant and quickly deteriorates at SD > 0.91. Frost resistance (F) is inversely related to the SD value:

$$F \sim \frac{1}{SD} = 1 + \frac{V_{e.a.}}{V_{fr.w.}} \qquad (9.4)$$

From the calculated SD values (Table 9.3), it follows that for fibre concrete with admixture of superplasticizer, it is possible to predict sufficiently high frost resistance. This prediction was confirmed by experiments. Frost resistance was obtained for 10×10×10 cm specimens using the method according to Ukrainian standard [85] with water saturation in a 5% sodium chloride solution on freezing to a temperature of (–50 ± 2)°C and thawing at (20 ± 2)°C. Frost resistance was evaluated by the change in cubic specimen strength after 4, 8, 15, 19 and 27 alternating freezing and thawing cycles, the number of which corresponded to the frost resistance of levels – F150, F300, F500, F600 and F800.

The frost resistance coefficient K_F, which is the ratio of the compressive strength of the specimens, after a certain number of alternating freezing and thawing cycles and the control specimens' strength, was used as frost resistance criterion. Table 9.4 presents the number of cycles corresponding to the frost resistance level and the value of the frost resistance coefficient.

Figures 9.6 and 9.7 show change in cubic strength during testing.

As can be seen from the data in Table 9.4, the obtained frost resistance values correlate well enough with the calculated SD criterion ones. At the same time, the

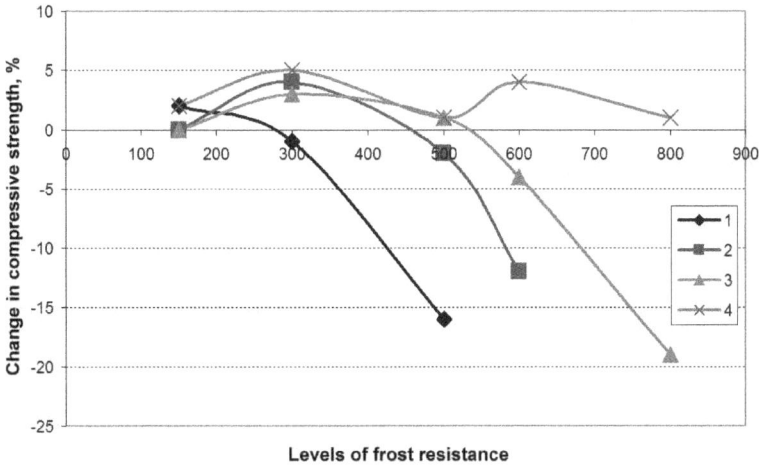

Fig. 9.6: Dependencies of strength change for ordinary and fibre-reinforced concrete specimens vs. the number of freezing-thawing cycles (composition numbers are according to Table 7.15)

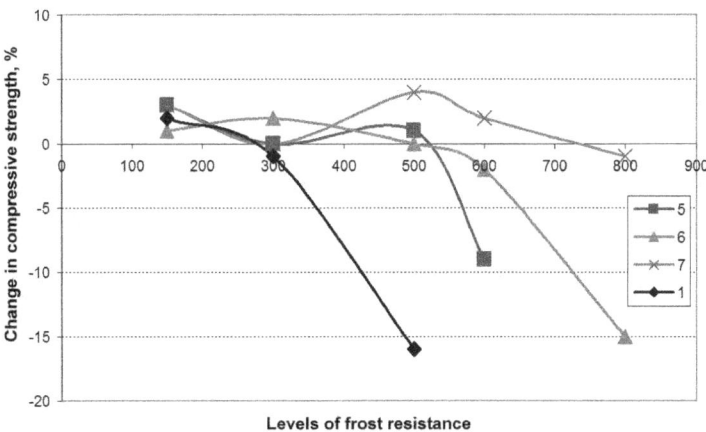

Fig. 9.7: Dependencies of strength change of fine-grained concrete and fibre-reinforced concrete specimens vs. the number of freezing-thawing cycles (composition numbers are according to Table 7.15)

fibre presence factor, which is well observed in increased frost resistance values in compositions 2 and 6 as compared to 1 and 5, respectively, is traceable.

Table 9.4: Concrete frost resistance

W/C	Fibre, kg/m³	Plasticizer type and content, %	Number of cycles by the third method	Frost resistance level	Frost resistance coefficient
			Normal-weight Concrete		
0.46	–	–	8	300	0.95
0.48	60	–	15	500	0.98
0.36	60	C-3 (1%)	19	600	0.94
0.27	60	Melflux (0.5%)	27	800	1.01
			Fine-grained Concrete		
0.53	–	–	8	300	0.96
0.50	100	-	15	500	0.98
0.38	100	C-3 (1%)	19	600	0.96
0.32	100	Melflux (0.5%)	27	800	0.99

It should be noted that there were fluctuations in the specimens' strength during the tests (the frost resistance coefficient) both downward and upward (Figs. 9.6 and 9.7). This can be explained by filling concrete pores by salt crystals from the solution, with which the specimens were saturated during the tests.

Thus, experimental studies on composition factors effect on the steel-reinforced concrete frost resistance demonstrate the opportunity to achieve frost resistance F800 and potentially even more.

Residual Flexural Strength

Along with compressive and flexural strength, one of the main normalised strength indicators of fibre-reinforced concrete is the residual flexural strength. In accordance with this indicator, the fibre-reinforced concrete class is obtained according to the residual flexural strength B_F. Introducing the B_F class allows to correctly take into account the nature of the material's operation after the cracks formation, which was not taken into account by previous codes [80].

Determining the B_F class for a specific fibre-reinforced concrete composition is reduced to conducting a series of tests for 150×150×600 beam specimens in bending according to the four-point loading scheme (Fig. 9.8).

A feature of these tests is the need to construct a 'load deformation' or 'load-crack opening' diagram with a continuous recording mode during the tests for allowing qualitative and quantitative assessment of material performance after

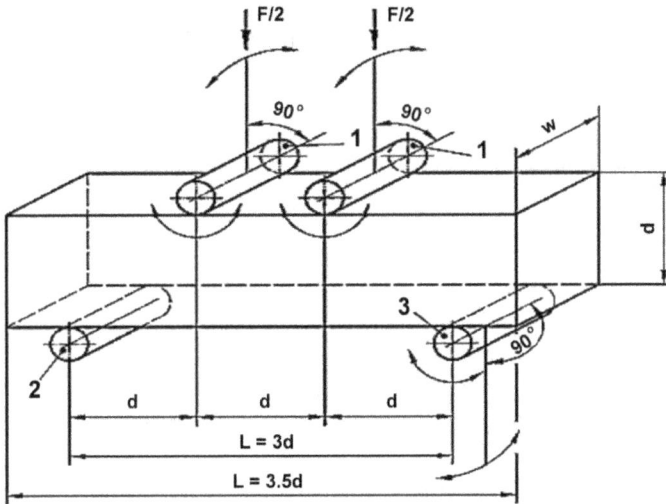

Fig. 9.8: The beam specimen test scheme: 1 – loading roller; 2 – supporting roller; 3 – supporting roller (with free rotation and tilt)

cracks formation. During the tests, the following values are determined: maximum load preceding the first crack opening (proportionality limit) F_L; load value for small deformations range $F_{0.5}$ and ultimate deformations $F_{2.5}$.

When classifying fibre-reinforced concrete according to the residual flexural strength, the main strength characteristics are the value of their residual flexural strength for the small deformations range $f_{F0.5}$ and ultimate deformations $f_{F2.5}$ (with a guaranteed security of 0.95). The small deformations range corresponds to crack opening width of 0.5 mm for testing beam specimens with a notch or a deflection of 0.75 mm (1/600 of the tested specimen span) when testing beam specimens without a notch. The ultimate deformations' range corresponds to the crack opening width of 2.5 mm when testing beam specimens with a notch, or a deflection of 3.0 mm (1/150 of the tested specimen span) when testing beam specimens without a notch.

Class B_F is indicated by a number and a letter. The number characterises the strength $f_{F0.5}$ with rounding to the smaller side with a multiple of 0.5 MPa, while the latter specifies the ratio $f_{F2.5}/f_{F0.5}$:

'a' – at $0.5 \leq f_{F2.5}/f_{F0.5} < 0.7$, 'b' – at $0.7 \leq f_{F2.5}/f_{F0.5} < 0.9$, 'c' – at $0.9 \leq f_{F2.5}/f_{F0.5} < 1.1$, 'd' – at $1.1 \leq f_{F2.5}/f_{F0.5} < 1.3$ and 'e' – at $1.3 \leq f_{F2.5}/f_{F0.5}$.

For example, the fibre-reinforced concrete class by the residual flexural strength for fine-grained fibre-reinforced concrete (composition No. 8, Table 7.15) is obtained as follows. Three specimens of fine-grained fibre-reinforced concrete beams were used. The beams were produced and hardened under normal conditions for 28 days. The load-deformation diagrams for three specimens are shown in Fig. 9.9. The values of loads $F_{0.5}$ and $F_{2.5}$ for different material operation stages and corresponding values of the residual flexural of fibre-reinforced concrete strengths $f_{F0.5}$ and $f_{F2.5}$ are given in Table 9.5.

Table 9.5: Residual flexural strength

Specimen number	F_L, N	$F_{0.5}$, N	$F_{2.5}$, N	$f_{F0.5}$, MPa	$f_{F2.5}$, MPa
1	120000	56250	43500	7.50	5.80
2	114000	53438	41325	7.13	5.51
3	122250	64500	46500	8.60	6.20

The values of $f_{F0.5}$ and $f_{F2.5}$ in this case (four-point bending of beam specimens without a notch) are calculated by using the following equation:

$$R_{Fi} = \frac{F_i \cdot l_i}{b_i \cdot h_i^2} \tag{9.5}$$

where l_i is i-th beam specimen span (450 mm); b_i and h_i are width (150 mm) and height (150 mm) of the i-th beam specimen cross-section, respectively.

After statistical processing of the $f_{F0.5}$ and $f_{F2.5}$ values, it is possible to obtain the fibre-reinforced concrete class based on the residual flexural strength B_F. Average strength values:

$$f_{F0.5,\,m} = 7.74 \text{ MPa}; f_{F2.5,\,m} = 5.84 \text{ MPa}$$

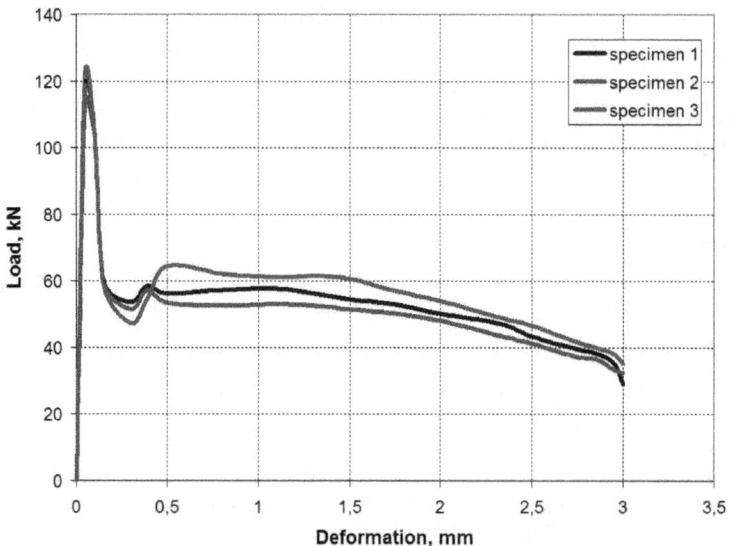

Fig. 9.9: Load-deformation diagramss for the investigated specimens of fine-grained fibre-reinforced concrete

The value of mean square deviations:

$$S_{F0.5, m} = 0.7666; S_{F2.5, m} = 0.3465.$$

Variation coefficients:

$$V_{F0.5, m} = 0.0990; V_{F2.5, m} = 0.0593.$$

Strength value with confidence probability 0.95:

$$f_{F0.5} = 6.30 \text{ MPa}; f_{F2.5} = 5.18 \text{ MPa}.$$

The ratio $f_{F2.5}/f_{F0.5}$ in this case is 5.18/6.30 = 0.82.

Thus, this fine-grained fibre-reinforced concrete corresponds to the class in terms of residual flexural strength B_F6b.

High-strength Fibre-reinforced Concrete with Composite Dispersed Reinforcement

To study the possibility of obtaining polydisperse fibre-reinforced concrete using steel and basalt or polypropylene fibres and providing their distribution uniformity in the resulting structure, a series of corresponding fibre-reinforced concrete specimens were prepared. The following composition of fine-grained concrete was used as the basic one: Portland cement – 500 kg/m^3, the aggregate ratio (crushed stone fraction 2-5 mm : sand = 0.55 : 0.45) to cement was 3.6 to 1 (by weight). The mixture W/C was 0.35. The required mixture slump of 130-150 mm was achieved by adding the Melflux 2651f superplasticizer. Steel corrugated fibre F1, 12 and 24 mm long basalt fibre and 12 and 18 mm long polypropylene fibre were used. At the same time, the steel fibre content varied between 80 and 120 kg/m^3, basalt fibre – from 0 to 6 kg/m^3 and polypropylene – from 0 to 2 kg/m^3.

Fibre-reinforced concrete with composite dispersed reinforcement was produced as follows. Basalt or polypropylene fibre was added to the plasticizer solution and mixed in a laboratory mixer with a vertical shaft for 40-60 sec. Cement was added to the fluff. In this way, fibre and mixing was again carried out until a homogeneous suspension was obtained. Then aggregate was added to the resulting suspension. Finally, in the last stage, at continuous mixture mixing, the necessary steel fibre content sifted through a sieve was added. This technology allowed avoidance of fibres' clumping and ensuring the necessary homogeneity of composite dispersed-reinforced concrete.

Effect of Metallic and Non-metallic Fibre Content and Ratio on Fibre-reinforced Concrete Strength

Table 10.1 presents the obtained strength values of fibre-reinforced concrete based on steel and basalt fibre with different combinations of polydisperse reinforcement.

As it follows from the data in Table 10.1, adding basalt fibre has a positive effect on concrete flexural strength. The fibre-reinforced concrete strength $f_{c,tf}^7$ for the control composition at a steel fibre content of 80 kg/m^3 was 10.3 MPa, and its maximum value for composite fibre concrete with basalt fibre was 12.4 MPa (at fibre content of 4 kg/m^3 and a fibre length of 12 mm). The average increase in flexural strength, depending on fibre length and its content, was from 7 to 20%.

With steel fibre content of 120 kg/m^3, the strength of the control composite fibre-reinforced concrete $f_{c,tf}^7$ was 13.3 MPa, and its maximum value for composite fibre-reinforced concrete reached 14.9 MPa (with a basalt fibre content of 4 kg/m^3 and fibre length of 12 mm). Thus, the maximum increase in strength was 12%. Basalt fibre with a length of 24 mm generally showed worse results. The increase in $f_{c,tf}^7$ was just by 7% at fibre content of 2 kg/m^3 (steel fibre content was 80 kg/m^3). Increase in basalt fibre content yielded a maximum decrease in strength of 40%.

The decrease in fibre-reinforced concrete strength, observed when the fibre length increased, is probably caused by a decrease in fibre percentage in the

Table 10.1: Strength of fibre-reinforced concrete specimens with different content and ratio of steel and basalt fibre

Steel fibre content, kg/m^3	Basalt fibre content, kg/m^3	Total fibre content by volume, μ, %	Part of basalt fibre in the total reinforcement volume, n	Compressive strength at seven days f_{cm}^7, MPa	Flexural strength at seven days $f_{c,tf}^7$, MPa
			Basalt Fibre (l = 12 mm)		
80	0	1.03	0.00	64.5	10.3
	2	1.10	0.07	65.2	12.1
	4	1.18	0.13	67.1	12.4
	6	1.25	0.18	66.2	10.2
120	0	1.54	0.00	66.8	13.3
	2	1.61	0.05	66.5	14.8
	4	1.69	0.09	69.2	14.9
	6	1.76	0.13	65.4	11.5
			Basalt Fibre (l = 24 mm)		
80	0	1.03	0.00	64.5	10.3
	2	1.10	0.07	67.2	11
	4	1.18	0.13	63.1	9.6
	6	1.25	0.18	62.8	6.54
120	0	1.54	0.00	66.8	13.3
	2	1.61	0.05	68.6	13.45
	4	1.69	0.09	63.1	11.94
	6	1.76	0.13	61.7	7.91

fracture section and a decrease in the bond strength at the fibre-cement matrix border. Additionally, fibres with a length of 24 mm are distributed worse in the fibre-reinforced concrete mixture, which affects the features of its structure.

Analysis of the destructive nature of composite fibre-reinforced concrete with 12 mm long basalt fibres shows that concrete destruction occurs with a fairly good fibres' participation in the cement matrix behaviour. Considering the effect of the basalt fibre part in the total reinforcement volume (Fig. 10.1), it should be noted that its optimal content is 2 kg/m^3 ($n = 0.07$) at steel fibre content of 80 kg/m^3 and 2-4 kg/m^3 ($n = 0.05$-0.09) at steel fibre content of 120 kg/m^3. Further increase in basalt fibre content obviously leads to higher specific surface area of dispersed reinforcement, which in turn leads to increase in the amount of water required to obtain concrete mixtures with a given workability.

Compressive strength of composite fibre-reinforced concrete, as expected, has little dependence on the basalt fibre content. For different compositions, the strength value changed on an average within 2-3%.

To compare the results obtained for basalt fibre, experiments were carried out, using polypropylene fibre, which belongs to the low-modulus fibres group. As can be seen from the results given in Table 2 for various combinations of steel and polypropylene fibre consumption, in general, there is a deterioration in flexural strength values in comparison with the control specimens. Only for steel 12 mm length polypropylene fibre and its content of 1 kg/m^3, there was almost no decrease in the composite fibre-reinforced concrete strength $f_{c,tf}^7$. In all other cases, use of polypropylene fibre worsened the flexural strength by an average of 5-40%. The compressive strength practically did not change for different compositions within the investigated limits.

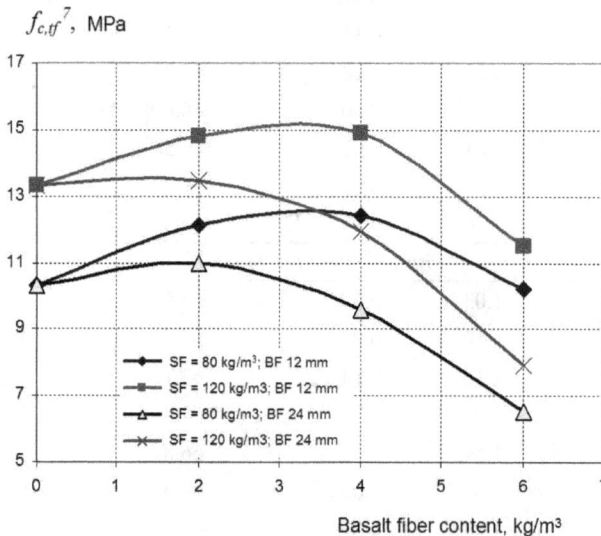

Fig. 10.1: Dependencies of flexural strength for polydisperse fibre-reinforced concrete

Analysing the obtained results, shown in Tables 10.1 and 10.2, it can be concluded that the use of polypropylene fibre as a component of composite reinforcement gives a slightly worse effect in comparison with basalt one. This can primarily be attributed to the poorer distribution of polypropylene fibres in the concrete structure, which does not contribute to uniform steel fibres' distribution.

The structure of destroyed fibre-reinforced concrete specimens with polypropylene fibre indicated signs of structural inhomogeneity, fibre lumps were observed in individual zones and such mixtures were more prone to delamination during casting and moulding. Examination of the destruction nature composite of fibre-reinforced concrete specimens with polypropylene fibre indicated a worse adhesion of polypropylene to the cement matrix, as a result of which individual fibres were pulled out at the specimen failure.

All of the above-mentioned disadvantages were practically absent in the case of using basalt fibre for composite dispersed reinforcement. It is obvious that basalt fibre, which in terms of density and elastic characteristics is closer to the concrete matrix than polypropylene, at optimal content and ratio with steel fibre forms a kind of supporting framework that holds the steel fibre, prevents the mixture delamination and contributes to uniform distribution of dispersed reinforcement in the concrete structure. All these factors, accordingly, affect the fibre-reinforced concrete strength characteristics.

Table 10.2: Strength of fibre-reinforced concrete specimens with different content and ratio of steel and polypropylene fibre

Steel fibre content, kg/m^3	Polypropylene fibre content, kg/m^3	Total fibre content by volume, μ, %	Part of polypropylene fibre in the total reinforcement volume, n	Compressive strength at seven days f_{cm}^7, MPa	Flexural strength at seven days $f_{c,tf}^7$, MPa
Polypropylene Fibre (l = 12 mm)					
80	0	1.03	0.00	64.5	10.3
	1	1.14	0.10	64.2	10.1
	2	1.25	0.18	64.6	9.4
120	0	1.54	0.00	66.8	13.3
	1	1.65	0.07	67.2	12.7
	2	1.76	0.13	67.3	12.1
Polypropylene Fibre (l = 18 mm)					
80	0	1.03	0.00	64.5	10.3
	1	1.14	0.03	63.2	9.8
	2	1.25	0.06	62.9	8.6
120	0	1.54	0.00	66.8	13.3
	1	1.65	0.07	65.3	10.6
	2	1.76	0.13	65.9	7.8

The Influence of Composition Factors on Strength of Fine-grained Fibre-reinforced Concrete with Composite Dispersed Reinforcement

Previous studies confirm the hypothesis about the possibility of improving the structure and, accordingly, the fibre-reinforced concrete properties by composite dispersed reinforcement when using basalt fibre.

In order to study this issue in detail, at the next stage, the complex influence of cement consumption, water-cement ratio, steel and basalt fibre content and volume ratio on strength characteristics of fine-grained fibre-reinforced concrete with composite dispersed reinforcement was investigated. With this aim, a three-level, four-factor B_4 experiment plan was implemented [42].

The raw materials used were Portland cement CEM-I with 28-day compressive strength 50 MPa, aggregate fraction mixture of 0.16-2 mm (quartz sand with M_f = 2.1) and 2-5 mm (crushed granite stone) in the ratio of 0.45 : 0.55. The necessary mixture workability (Sl = 130-150 mm) was achieved by adding the Melflux 2651f superplasticizer. Steel corrugated fibre F1 and 12 mm long basalt fibre were used. The experiment planning conditions and the obtained results are given in Tables 10.3, 10.4 and 10.5 respectively.

After the experimental data processing and statistical analysis, mathematical models for compressive and flexural strengths of fine-grained fibre concrete with composite dispersed reinforcement were obtained in the form of polynomial regression equations. The results of experimental data processing and statistical analysis are shown in Table 10.6.

Using the models given in Table 10.6 graphical dependencies and response surfaces of initial parameters vs. two influencing factors were constructed (Figs. 10.2 and 10.3). At the same time, two other factors, not represented in each of the graphs, were fixed at the zero level.

Table 10.3: Experiment planning conditions for studying composition parameters of fibre-reinforced concrete with composite dispersed reinforcement

Number of factors	Factors		Variation levels			Variation interval
	Coded	Natural form	−1	0	+1	
1	X_1	Cement consumption, kg/m³ (C)	450	500	550	50
2	X_2	W/C	0.3	0.35	0.4	0.05
3	X_3	Steel fibre content, kg/m³ (SF)	80	100	120	20
4	X_4	Basalt fibre content, kg/m³ (BF)	0	2	4	2

Analysing the obtained experimental-statistical models for compressive strength, it can be noted that the expected most significant factor affecting it is the water-cement ratio (X_2), which decreases from +1 to −1 (from W/C = 0.4 to W/C

Table 10.4: Matrix of experiments planning for investigating the polydisperse reinforcement effect on fine-grained concrete strength

Plan point number	Coded values of factors				Components consumption, kg/m³					
	X_1	X_2	X_3	X_4	C	S	A	W	SF	BF
1	+	+	+	+	550	732	895	220	120	4
2	+	+	+	-	550	732	895	220	120	0
3	+	+	-	+	550	732	895	220	80	4
4	+	+	-	-	550	732	895	220	80	0
5	+	-	+	+	550	799	977	165	120	4
6	+	-	+	-	550	799	977	165	120	0
7	+	-	-	+	550	799	977	165	80	4
8	+	-	-	-	550	799	977	165	80	0
9	-	+	+	+	450	820	1002	180	120	4
10	-	+	+	-	450	820	1002	180	120	0
11	-	+	-	+	450	820	1002	180	80	4
12	-	+	-	-	450	820	1002	180	80	0
13	-	-	+	+	450	875	1069	135	120	4
14	-	-	+	-	450	875	1069	135	120	0
15	-	-	-	+	450	875	1069	135	80	4
16	-	-	-	-	450	875	1069	135	80	0
17	+	0	0	0	550	766	936	193	100	2
18	-	0	0	0	450	847	1036	158	100	2
19	0	+	0	0	500	776	948	200	100	2
20	0	-	0	0	500	837	1023	150	100	2
21	0	0	+	0	500	806	986	175	120	2
22	0	0	-	0	500	806	986	175	80	2
23	0	0	0	+	500	806	986	175	100	4
24	0	0	0	-	500	806	986	175	100	0

= 0.3) and leads to an average increase in strength of 40%. Moreover, such an increase is observed at different cement consumptions and at different hardening durations.

By the influence on compressive strength of fine-grained fibre-reinforced concrete with composite dispersed reinforcement, the investigated factors can be arranged in the following order: $X_2 > X_1 > X_4 > X_3$.

The nature of the dependencies for flexural strength, obtained by using corresponding models (Table 10.6), is significantly different. It should be noted that the steel fibre content (X_3), which reflects the dispersed reinforcement degree in fibre-reinforced concrete, is one of the most significant factors in this strength.

Moreover, the maximum effect of this factor is manifested at 28 days (38% of all factors effect). W/C has the second highest influence degree on flexural strength (33%), and its influence at an early age even outweighs the influence of the steel fibre content (Fig. 10.4).

The influence of this factor in the selected variation range is linear (Fig. 10.2) and it makes up about 70% of the influence of all factors. Increasing cement consumption (X_1) within the variation limits increases the compressive strength by 8-20%. The change in steel fibre content (X_3) at a constant water-cement ratio does not significantly affect the tested concrete strength. At the same time, basalt fibres (X_4) contribute a slight increase in fibre-reinforced concrete strength, especially at 28 days (Fig. 10.3). This is also indicated by the values of linear coefficients in regression equations (–0.1 for X_3 and 1.0 for X_4). Such a difference in the influence of steel and basalt fibre on fibre-reinforced concrete strength can be explained by higher specific surface of the latter and its better adhesion to the concrete matrix.

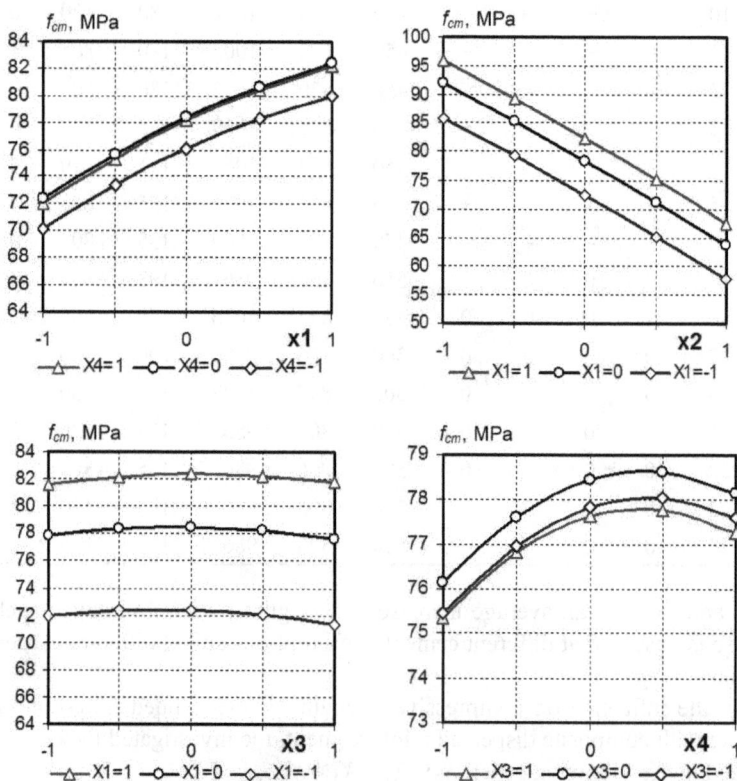

Fig. 10.2: Dependencies of compressive strength for fibre-reinforced concrete with composite dispersed reinforcement at 28 days on cement consumption (X_1), W/C (X_2), steel and basalt fibre contents (X_3 and X_4)

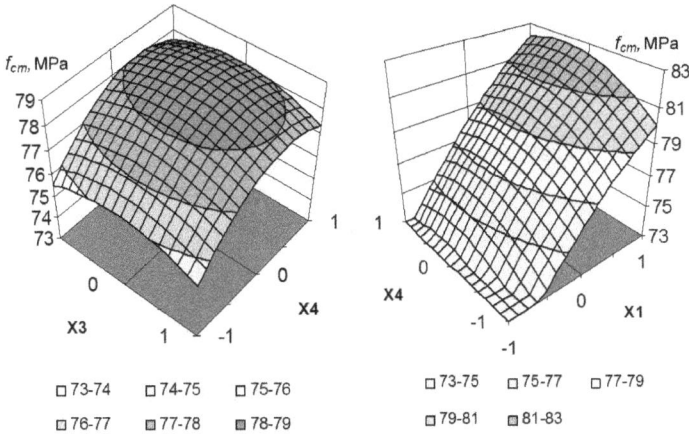

Fig. 10.3: Response surfaces of compressive strength at 28 days for fibre-reinforced concrete with composite dispersed reinforcement vs. cement consumption (X_1), steel and basalt fibre contents (X_3 and X_4)

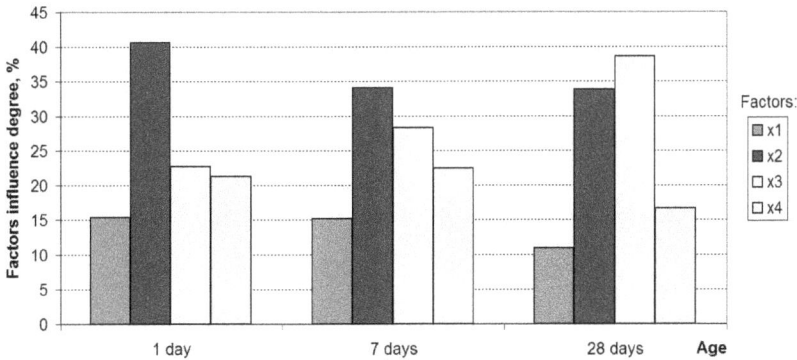

Fig. 10.4: Factors influence degree on flexural strength at different age

The effect of steel fibre content (X_3) has a weakly expressed extreme character, which can be seen both from the corresponding graphical dependence (Fig. 10.5) and from the quadratic coefficient in the corresponding regression equation (Table 10.6). An increase in fibre content from 80 ($X_3 = -1$) to 110 ($X_3 = 0.5$) kg/m³ leads to an increase in flexural strength by an average of 30-40%, depending on the values of other factors.

A further increase in the steel fibre content does not significantly affect the fine-grained fibre-reinforced concrete flexural strength. The lowest effect on the considered indicator at a constant W/C has the cement consumption factor (X_1), which has the smallest linear coefficient in the regression equation. A relatively high interaction coefficient of factors X_1 and X_2 indicates a significant dependence of the influence degree of each of them on the change of the other and leads to ambiguous curves on the graph (Fig. 10.5).

Table 10.5: Experimental composition parameters and tests results

Plan point number	W/C	SP, %	f_{cm}^{1}, MPa	$f_{c,tf}^{1}$, MPa	f_{cm}^{7}, MPa	$f_{c,tf}^{7}$, MPa	f_{cm}^{28}, MPa	$f_{c,tf}^{28}$, MPa
1	0.4	0.25	27.4	7.1	55.2	14.2	64.9	16.7
2	0.4	0.20	28.6	6.2	54.7	11.1	67.2	13.8
3	0.4	0.13	30.2	6.3	57.1	12.4	67.3	12.1
4	0.4	0.05	28.2	5.0	55.8	10.0	66.6	9.0
5	0.3	1.18	40.5	9.4	76.9	17.9	95.9	20.9
6	0.3	1.10	40.9	8.8	78.1	15.9	96.0	19.9
7	0.3	0.87	41.6	8.9	80.3	18.3	96.4	17.8
8	0.3	0.80	40.3	7.7	79.7	15.5	95.2	15.2
9	0.4	0.44	24.1	7.2	49.2	15.2	55.9	16.2
10	0.4	0.40	23.8	6.6	49.1	13.9	57.3	14.9
11	0.4	0.26	24.5	6.1	49.6	10.8	57.1	11.4
12	0.4	0.20	23.0	5.2	48.0	8.2	56.9	9.5
13	0.3	1.14	33.9	8.3	69.9	16.9	83.8	19.8
14	0.3	1.05	34.1	7.5	71.0	15.2	83.9	17.0
15	0.3	0.34	34.3	7.2	72.2	12.8	84.8	13.7
16	0.3	1.00	33.8	6.3	70.4	10.3	83.3	11.8
17	0.35	0.48	32.3	8.3	67.1	16.9	79.9	19.2
18	0.35	0.32	29.9	7.6	64.2	15.6	75.1	18.3
19	0.4	0.33	24.2	7.7	53.0	14.7	65.0	16.3
20	0.3	0.56	39.5	8.8	77.9	17.4	92.9	20.1
21	0.35	0.34	31.7	7.5	68.1	14.2	77.2	16.4
22	0.35	0.13	32.3	6.8	62.3	11.5	78.5	12.1
23	0.35	0.27	32.0	8.7	65.2	15.7	76.4	18.9
24	0.35	0.20	31.6	7.5	66.0	14.6	78.1	17.1

Regarding the effect of the basalt fibre content (X_4), as already mentioned earlier, its addition into the mixture composition of up to 4 kg/m³ allows to increase the flexural strength by up to 20% as compared to compositions without basalt fibre. It is also possible to note the high influence degree of this factor on the strength value at the early hardening stages, which at one day is more than 20% (Fig. 10.4) and is almost not inferior to the influence of the steel fibre

Table 10.6: Mathematical models of strength parameters for fine-grained fibre-reinforced concrete with composite dispersed reinforcement

Output parameter		Mathematical models0
Compressive strength:	One day	$f_{cm}^1 = 31.7 + 2.7X_1 - 5.8X_2 - 0.2X_3 + 0.2X_4 - 0.6X_1^2 +$ $0.1X_2^2 + 0.3X_2^3 - 0.5X_1X_2 - 0.3X_1X_3 - 0.4X_3X_4$ (10.1)
	Seven days	$f_{cm}^7 = 66.1 + 3.6X_1 - 11.3X_2 - 0.2X_3 + 0.2X_4 - 0.7X_2^2 -$ $2.4X_4^2 - 0.4X_1X_2 - 0.5X_1X_3 + 0.2X_2X_3 + 0.3X_2X_4 - 0.4X_3X_4$ (10.2)
	28 days	$f_{cm}^{28} = 78.4 + 5X_1 - 14.2X_2 - 0.1X_3 + 1X_4 - 1.1X_1^2 - 0.6X_2^2 -$ $0.7X_3^2 - 1.3X_4^2 - 0.1X_1X_2 + 0.2X_1X_3 - 0.35X_2X_3$ (10.3)
Flexural strength:	One day	$f_{c,tf}^1 = 7.47 + 0.31X_1 - 0.82X_2 + 0.46X_3 + 0.43X_4 - 0.18X_1^2 -$ $0.33X_3^2 + 0.12X_4^2 - 0.36X_1X_2 - 0.1X_1X_4 - 0.1X_3X_4$ (10.4)
	Seven days	$f_{c,tf}^7 = 15.56 + 0.74X_1 - 0.66X_2 + 1.38X_3 + 1.1X_4 + 0.67X_1^2 +$ $0.47X_2^2 - 1.43X_3^2 - 0.43X_4^2 - 0.8X_1X_2 - X_1X_3 + 0.14X_1X_4 +$ $0.25X_2X_3 - 0.14X_3X_4$ (10.5)
	28 days	$f_{c,tf}^{28} = 17.85 + 0.66X_1 - 2.03X_2 + 2.32X_3 + X_4 +$ $0.88X_1^2 + 0.33X_2^2 - 1.62X_3^2 - 0.57X_4^2 - 0.75X_1X_2 -$ $0.18X_1X_3 - 0.1X_1X_4 - 0.19X_3X_4$ (10.6)
Melflux 2651F content		$SP = 0.41 + 0.095X_1 - 0.33X_2 + 0.12X_3 + 0.11X_4 - 0.06X_1^2 +$ $0.2X_2^2 - 0.03X_3^2 - 0.02X_4^2 - 0.07X_1X_2 - 0.02X_1X_3 +$ $0.04X_1X_4 - 0.05X_2X_3 + 0.04X_2X_4 + 0.04X_3X_4$ (10.7)

content (X_3). Subsequently, there is a decrease in the effect of factor X_4 and a corresponding increase in X_3.

Analysis of the response surfaces for normalised parameter $f_{c,tf}^{28}$ (Fig. 10.6) indicates that in order to achieve the maximum values of flexural strength, it is necessary to maintain the value of *W/C* factor (X_2) at the lower variation level, and the contents of steel (X_3) and basalt fibre (X_4) at the upper level. It can also be concluded that achieving values of $f_{c,tf}^{28} > 18$ MPa can be ensured in a wide range of steel and basalt fibre contents ($X_3 = 0$-1; $X_4 = -0.5$-1) – respectively 100-120 kg/m^3 of steel and 1-4 kg/m^3 of basalt fibre.

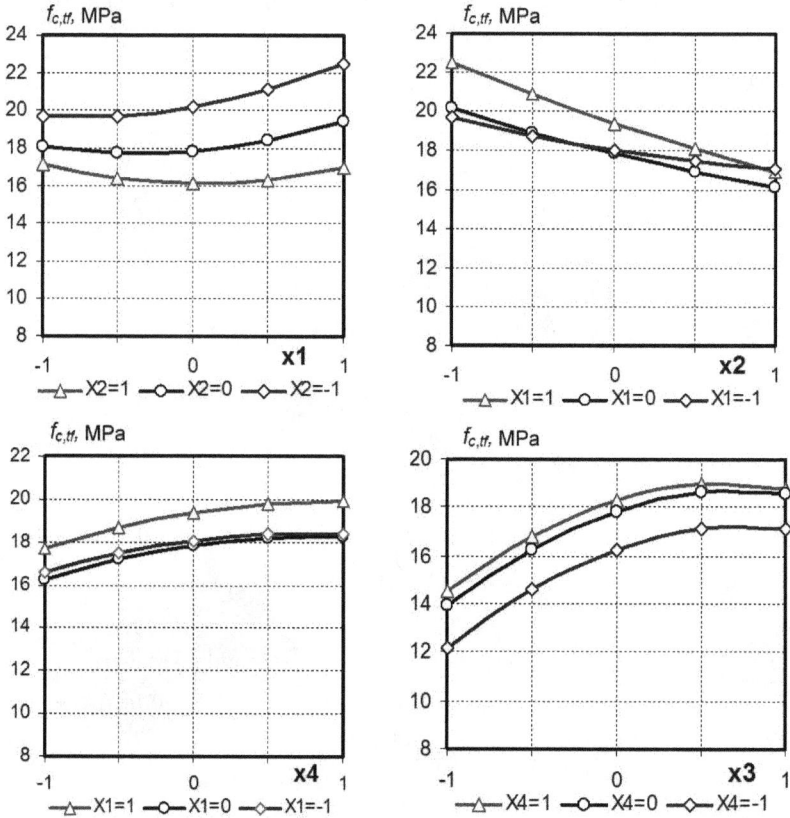

Fig. 10.5: Dependencies of flexural strength for fibre-reinforced concrete with composite dispersed reinforcement at 28 days on cement consumption (X_1), W/C (X_2), steel and basalt fibre contents (X_3 and X_4)

Analysis of the superplasticizer content model (Fig. 10.7) allows us to note the highest influence of W/C (X_2). The value of its linear coefficient in the regression equation (Table 10.6) significantly exceeds those for the other two factors. An increase in steel (X_3) and basalt fibre (X_4) contents within the variation limits leads to an increase in superplasticizer content, which is explained by the need to ensure the required workability with a potential increase in water demand. An increase in cement consumption in the range from 450 ($X_1 = -1$) to 500 kg/m³ ($X_1 = 0$) yields a necessary increase in superplasticizer content of 1.5-2 times; further ($X_1 = 0-+1$) this dependence significantly decreases.

It should be noted that despite some increase in the concrete mixture water demand when basalt fibre is used, which must be compensated by higher superplasticizer content, other indicators of both the mixture and hardened fine-grained concrete improve.

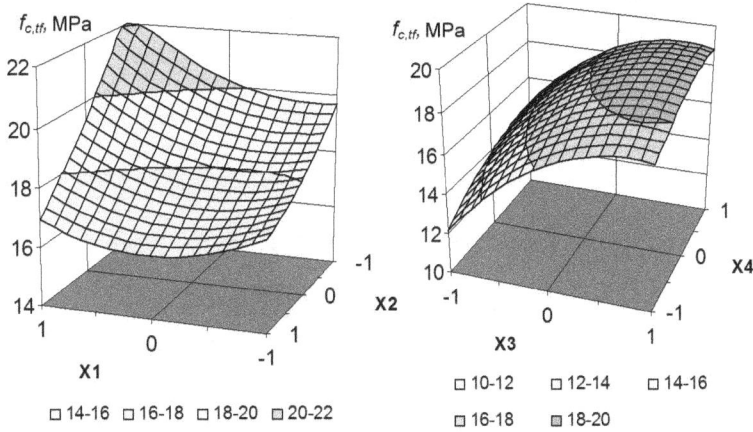

Fig. 10.6: Response surfaces of flexural strength at 28 days for fibre-reinforced concrete with composite dispersive reinforcement vs. cement consumption (X_1), W/C (X_2), steel fibre and basalt contents (X_3 and X_4)

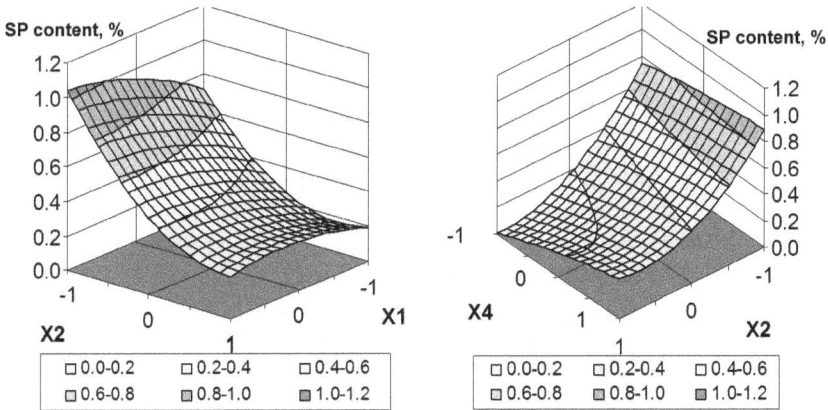

Fig. 10.7: Response surfaces of superplasticizer content for fibre-reinforced concrete with composite dispersed reinforcement vs. cement consumption (X_1), W/C (X_2) and basalt fibre content (X_4)

According to the results of previous studies, the optimal steel fibre content was about 100 kg/m³ ($\mu = 1.3\%$). A further increase in the fibre content led to a decrease in strength, which was the result of a tendency for delamination even under low loads. When using composite dispersed reinforcement with basalt fibre, it is possible to achieve the steel fibre content of 120 kg/m³, providing the structure uniformity and practically no delamination, and strength indicators of such fibre-reinforced concrete increase by 10-20%.

Dynamic Characteristics of Fibre-reinforced Concrete

Concrete is an elastic-plastic material that clearly exhibits plastic properties under static loads. At the same time, under dynamic loads, concrete can behave in different ways, namely, withstand rather high dynamic loadings, while increasing the static resistance to compression, tension or splitting, or collapse [86, 87]. It is believed that the shorter the dynamic loads' duration, the more the dynamic resistance exceeds the static one. This duration is tenths of a second, or several seconds; if the load does not decrease during this time, concrete will collapse. It should be noted that there are cases of a decrease in dynamic resistance of concrete compared to the static one [87, 88]. This is characteristic for brittle high strength concrete, which must be taken into account in compositions' design.

In general, dynamic characteristics of concrete and fibre-reinforced concrete include the crack resistance, impact strength and impact toughness [80, 87]. Crack resistance is determined by the ability to withstand impact loads without cracks formation. For fibre-reinforced concrete, it is common to determine strength characteristics of standard specimens without cracks and also with a certain opening width [89-91].

The Effect of Dispersed Reinforcement Factors on Impact Strength of Fibre-reinforced Concrete

Cube specimens with an edge of 10 cm were prepared to test high strength fibre-reinforced concrete and obtain its impact strength. As reinforcing elements, corrugated steel fibres of 60, 40 and 20 mm length and basalt fibres of 12 mm length were used. The impact strength was estimated by the amount of work (A) spent on destroying the specimen, related to its volume. Table 11.1 and Fig. 11.1 present the results of determining the work required for a fibre-reinforced concrete specimen destruction, depending on the type, content and length of the reinforcement.

Table 11.1: The work required for destruction of fibre-reinforced concrete specimens depending on fibre length and content

Steel fibre content (l = 60 mm), kg/m³	Basalt fibre content (l = 12 mm), kg/m³	Total fibre content by volume, μ, %	Part of basalt fibre in the total reinforcement volume, n, %	Work required for specimen destruction (A), J/cm³
-	-	-	-	9.2
Basalt Fibre (l = 12 mm)				
80	0	1.03	0.00	17
	2	1.10	0.07	19.8
	4	1.18	0.13	22.8
	6	1.25	0.18	20.2
120	0	1.54	0.00	20.2
	2	1.61	0.05	25.7
	4	1.69	0.09	31.5
	6	1.76	0.13	20.9

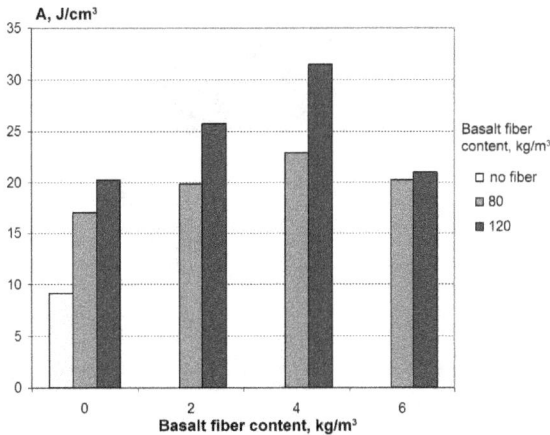

Fig. 11.1: Work required for fibre-reinforced concrete destruction depending on dispersed reinforcement type and content

As can be seen from the obtained results, adding dispersed reinforcing fibres into the fine-grained concrete composition significantly increases the work required to destroy the specimens. The maximum value of A is achieved with composite dispersed reinforcement – 31.5 J/cm³, which is 3.5 times more than for the control specimen without fibre. The earlier conclusion about the positive effect of basalt fibre on fibre-reinforced concrete properties is also confirmed. As can be seen from Fig. 11.1, an increase in basalt fibre content in polydisperse reinforcement leads to an increase in the destruction work and reaches its maximum at a content of 4 kg/m³. At steel fibre content of 80 kg/m³, the increase is 35%, and at 120 kg/m³, it is over 80%. Further increase of basalt fibre content is impractical and leads to a gradual decrease in impact strength, which is explained by the difficulty of fibres' uniform distribution in the concrete matrix.

It can also be noted that for fibre-reinforced concrete compositions with steel fibre content of 80 kg/m³ and basalt fibre content of 2 kg/m³, the same impact strength is achieved as with reinforcement by steel fibre only with a content of 120 kg/m³, and if the basalt fibre content is 4 kg/m³, it becomes bigger by 15%.

Through the achieved experiment, the relationship between the polydisperse fibre-reinforced concrete impact strength and the reinforcement parameters was established. Two variation factors were used: steel fibre content 120 ± 20 kg/m³ (x_1) and steel fibre length 40 ± 20 mm (x_2). In all cases, basalt fibre content was 4 kg/m³. As a result, a complete quadratic dependence of the destruction work (A) on the specified factors was obtained:

$$A = 29.5 - 1.3x_1 + 2.4x_2 - 6.8x_1^2 - 2.7x_2^2 + 2.4x_1x_2 \qquad (11.1)$$

Based on this model, a response surface (destruction work) vs. the investigated factors was obtained (Fig. 11.2).

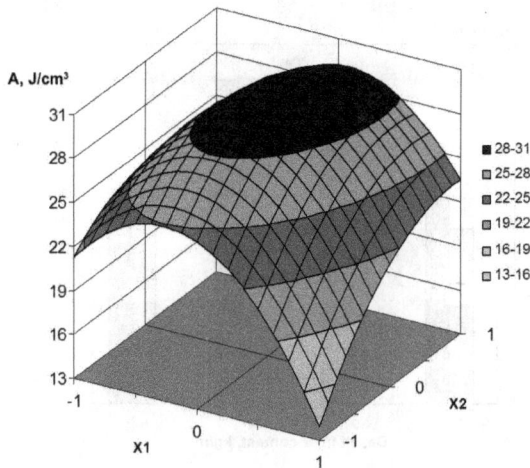

Fig. 11.2: Work for destruction of fibre-reinforced concrete specimens with composite dispersed reinforcement vs. steel fibre content (x_1) and fibres length (x_2) at basalt fibre content of 4 kg/m³

A decrease in the impact strength index of composite dispersed reinforced concrete with an increase in steel fibre content over 120 kg/m³ ($\mu = 1.5\%$) was obtained (Fig. 11.2). The maximum impact strength index was 31.8

J/cm^3, exceeding the control value for unreinforced concrete by more than 3.5 times. The first crack in the composite fibre-reinforced concrete was recorded after applying 36 impacts, whereas cracks formation in the control concrete composition occurred immediately after 14 impacts. The first crack opening width in the control composition concrete reached 1 mm and in the dispersed reinforced concrete, it was up to 0.2 mm.

The cracking resistance of fibre-reinforced concrete with 40 and 60 mm long fibres exceeds that with fibres 20 mm long. When using fibres longer than 40 mm, there is more overlap of one fibre with another, as a result of which, the cracks that form quickly disappear. Fibres with a length of 20 mm overlap to a lesser extent; therefore the opening cracks' length and width increase.

In general, the zone of optimal impact strength values is clearly visible in Fig. 11.2. It is observed when the steel fibres content is 110-125 kg/m^3 ($x_1 = -0.5$-0.25) and the fibres' length is 40-60 mm ($x_1 = 0$-1), the destruction work value will be at least 28 J/cm^3.

Impact Strength of Fibre-reinforced Concrete with Composite Dispersed Reinforcement

To test the impact strength of composite dispersed fibre-reinforced concrete 40×40×160 mm beams were prepared. F-1 corrugated steel fibre with a length of 60 mm and basalt fibres with a length of 12 mm were used as dispersed reinforcement. For experimental specimens production cement CEM-I with standard compressive strength 50 MPa, aggregate fractional mixture: 0.16-2 mm (quartz sand with $M_f = 2.1$) and 2-5 mm (crushed granite stone) in the ratio of 0.45 : 0.55 were used. The required mixture slump was achieved adding the Melflux 2651f superplasticiser. Cement consumption was 550 kg/m^3, $W/C = 0.35$.

The specimens used for the test had a square section with an 8 mm deep cut in the middle of the beams' span. Each tested specimen was placed tightly on the copra supports in such a way that the cut was located symmetrically to the supports and opposite to the applied impact. The destruction work for each specimen was determined by the pendulum deflection angle after impact or by the scale.

$$A = \frac{P \cdot L}{\cos\beta - \cos\alpha} \tag{11.2}$$

where P is the pendulum weight (N); L is the pendulum length – the distance from the axis to the centre of gravity, m; α and β are the pendulum rise angles before and after the specimen fracture, deg.

The impact strength IS [J/m^2], was obtained as:

$$IS = \frac{A_I}{S} \tag{11.3}$$

where A_I is the impact work, required for destruction, J; S is the cross-section area of the specimen at the cut point before testing, m^2.

The specimen destruction work values at impact bending by a pendulum copra are presented in Table 11.2. As follows from the obtained data, the value of the concrete specimen impact strength for the control composition was 1.16 J/m^2. The destruction work for dispersed reinforced concrete varied, depending on the reinforcing fibres' type and content. Adding 80 kg/m^3 of steel fibre leads to an increase in impact strength by six times (destruction work of 7.05 J/m^2), and for 120 kg/m^3, it was more than seven times (8.28 J/m^2). It was found that the destruction work for specimens with composite dispersed reinforcement increases by 30% on average and reaches a maximum value of 11.47 J/m^2 when the steel and basalt fibres' contents are 120 and 4 kg/m^3, respectively. As in the case of impact strength, it can be noted that combination of steel and basalt fibre in (80 and 4 kg/m^3 respectively) can replace mono-reinforced fibre concrete with a steel fibre content of 120 kg/m^3. The value of impact strength in both cases is practically the same.

Table 11.2: Work for destruction depending on the dispersed
reinforcement type and content

No.	Dispersed reinforcement type[*]	Fibre content, kg/m^3	Destruction work, J/m^2
1	–	–	1.16
2	SF	80	7.05
3	SF	120	8.28
4	SF + BF	80 + 4	10.14
5	SF + BF	120 + 4	11.47

[*] SF – steel fibre; BF – basalt fibre

Thus, dispersed reinforcement provides a significant increase in resistance to impact loads due to impact energy absorption by fibres. Concrete with such properties can be used in zones of high seismic activity for construction of road and airfield surfaces subjected to shock-type loadings.

Design of High-strength Fibre-reinforced Concrete Compositions

Composition design is one of the most important stages of concrete technology. Its properties, durability and cost-effectiveness depend on how well the concrete composition is determined. In this regard, serious attention is paid to development of new and improvement of existing methods for concrete composition design [24, 78, 79, 92-95]. The composition design method for a specific concrete type is developed, taking into account its specific features. As for fibre-reinforced concrete, there are none well-based or properly described in the literature rules, enabling to find its components ratio, taking into account its specific features and their interaction during the macrostructure formation as well as providing at the same time homogeneous fibre-reinforced concrete mixtures, required workability and hardened concrete strength and durability.

The fibre-reinforced concrete composition, macrostructure formation and its density are determined both by the concrete matrix components and by fibre reinforcement, its geometric parameters, orientation and saturation degree. Therefore, determining the composition of concrete itself, without taking into account the reinforcement parameters, as is common [75, 96], is one of the important reasons for the mixture quality deterioration when dispersive reinforcement is added into its composition and of the fibre-reinforcement effectiveness reduction.

In this study, based on experimental-statistical models, design methods for determining fibre-reinforced concrete compositions with composite dispersed reinforcement are proposed, which allow obtaining the required consumption of mixture components necessary to ensure a standardised set of construction and technical properties.

Fibre-reinforced Concrete Composition Design by the Nomographic Method

Nomograms of output parameters, based on experimental-statistical models, are often used to design compositions of various concrete types [42]. This is a typical management task, aimed at determining such combinations of factors that provide the required output parameters indicators [97]. With this aim, one of the factors is selected from the obtained regression equation, for example, flexural strength (Table 10.6). Solving the regression equation with respect to this factor, its necessary values, providing the given output parameter values when other factors change, are obtained. Figure 12.1 shows a nomogram for determining cement consumption at a given flexural strength of composite fibre-reinforced concrete. This nomogram in combination with previously obtained models (Table 10.6) can be used for design compositions of composite fibre-reinforced concrete with a set of specified properties.

Depending on the specific conditions, either water-cement ratio or contents of steel and basalt fibre can be the determining parameters of fibre-reinforced concrete composition. When designing compositions according to Table 12.1, we determine the desired range in which the composition of fibre-reinforced concrete with the specified compressive and flexural strength values can be found. Assuming certain fibre content or water-cement ratio according to the nomogram shown in Fig. 12.1, we determine the main parameters of the concrete mixture composition, which will ensure the specified flexural strength of concrete.

Table 12.1: Approximate values of composite fibre-reinforced concrete strength characteristics at 28 days

Basalt fibre content, kg/m³	Steel fibre content, kg/m³	W/C	$f_{c,tf}^{28}$, MPa	f_{cm}^{28}, MPa
0-2	80-100	0.3-0.35	14.2-17.8	70.3-95.3
		0.35-0.4	10.5-17.8	55.8-81.2
	100-120	0.3-0.35	16.7-20.95	68.9-94.4
		0.35-0.4	16.7-18.5	54.2-79.1
2-4	80-100	0.3-0.35	16.2-20.5	71.1-94.8
		0.35-0.4	12.1-17.17	55.9-82.0
	100-120	0.3-0.35	18.2-23.5	69.3-96.3
		0.35-0.4	17.8-18.8	55.8-79.7

Conversion of the fibre-reinforced concrete mixture composition parameters into a coded form is carried out according to the following dependencies:

$$x_1 = \frac{C-500}{50}; x_2 = \frac{W/C-0.35}{0.05}; x_3 = \frac{SF-100}{20}; x_3 = \frac{BF-2}{2} \qquad (12.1)$$

where C, SF, BF are cement consumption, steel and basalt fibre contents.

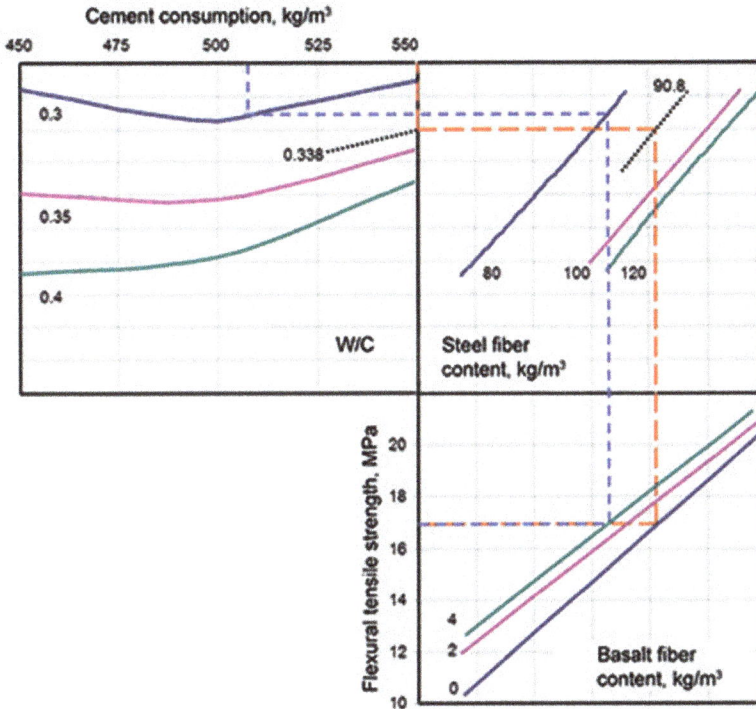

Fig. 12.1: Nomogram of flexural strength at 28 days for fibre-reinforced concrete with composite dispersed reinforcement

After substituting the obtained values into the equations (Table 10.6), we check whether the necessary concrete compressive and flexural strength at 28 days is ensured.

The water demand at the specified water-cement ratio and cement consumption is calculated according to the following dependence:

$$W = C \cdot W/C \qquad (12.2)$$

Substituting the coded values of cement consumption, fibre content and water-cement ratio into the equations (Table 10.6), the content of the polycarboxylate type superplasticizer, ensuring the necessary concrete mixture slump of 130-150 mm, is obtained. If it is necessary to ensure other concrete mixture workability, the superplasticizer content is specified experimentally.

With the obtained values of cement consumption and water demand, according to known methods [24], using Eqs. 12.3-12.6, the aggregates consumption is calculated, taking into account that the optimal ratio of sand and crushed stone of fraction 2-5 mm by weight is 45% and 55%, respectively.

The aggregate volume is:

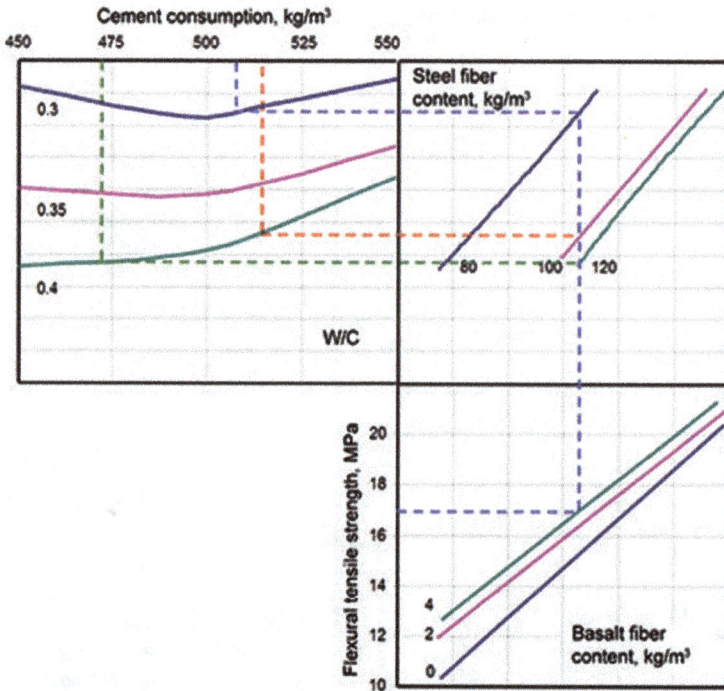

Fig. 12.2: Multivariate solution to the composition design problem for fibre-reinforced concrete with composite dispersed reinforcement by nomographic method

$$V_a = 1000 - \left(\frac{C}{\rho_c} + \frac{W}{\rho_w} + \frac{SF}{\rho_{SF}} + \frac{BF}{\rho_{BF}} \right) \tag{12.3}$$

The aggregate weight (sand + crushed stone):

$$m_a = V_a \cdot \rho_a \tag{12.4}$$

$$m_s = m_a \cdot 0.45 \tag{12.5}$$

$$m_{c.s} = m_a \cdot 0.55 \tag{12.6}$$

where ρ_c, ρ_a, ρ_{cf}, ρ_{bf} are real density of cement, aggregate, steel and basalt fibre; $V_{c.p}$ is the cement paste volume.

Example 1: *Calculate the composition of fine-grained fibre-reinforced concrete with composite dispersed reinforcement and a 28-day compressive strength of 70 MPa and flexural strength of 17 Mpa; the density of the aggregate mixture (sand and crushed granite stone fractions of 2-5 mm) $\rho_a = 2.7\ g/cm^3$.*

1. According to Table 12.1, the range of *W/C* and fibre content, in which the composition of fibre-reinforced concrete with specified values of

compressive and flexural strength can be found, is wide: steel fibre content is 80-120 kg/m³, basalt – 0-4 kg/m³ and water-cement ratio – 0.3-0.4.

2. According to the nomogram (Fig. 12.2), starting from the economy viewpoint with a minimum of steel fibre content of 80 kg/m³, the necessary cement consumption and water-cement ratio, which would ensure the required fibre-reinforced concrete flexural strength, is obtained.

3. Converting the obtained values ($C = 507$ kg/m³, $W/C = 0.3$, $SF = 80$ kg/m³, $BF = 4$ kg/m³) into coded form yields:

$$X_1 = \frac{(C-500)}{50} = \frac{(520-500)}{50} = 0.13$$

$$X_2 = \frac{(W/C-0.35)}{0.05} = \frac{(0.3-0.35)}{0.05} = -1$$

$$X_3 = \frac{(CF-100)}{20} = \frac{(80-100)}{20} = -1$$

$$X_4 = \frac{(BF-2)}{2} = \frac{(4-2)}{2} = 1$$

4. Substitute the obtained values into the equation from Table 10.6 and check whether the required concrete strength at 28 days is 75 MPa.

$f_{cm} = 78.4+5\cdot0.13\text{-}14.2\cdot(-1)\text{-}0.1\cdot(-1)+1\cdot1\text{-}1.1\cdot(0.13)^2\text{-}0.6\cdot(-1)^2\text{-}0.7\cdot(-1)^2\text{-}$
$1.3\cdot1^2\text{-}0.1\cdot0.13\cdot(-1)+0.2\cdot0.13\cdot(-1)\text{-}0.35\cdot(-1)\cdot(-1) = 91.4$ MPa.

The condition is fulfilled: $94.1 \geq 75$ MPa.

5. The water demand for a given water-cement ratio and cement consumption is found according to the equation:

$$W = C\cdot W/C = 507\cdot0.3 = 152 \text{ l/m}^3.$$

6. Substitution of cement consumption ($X_1 = 0.13$), water-cement ratio ($X_2 = -1$), steel fibre ($X_3 = -1$) and basalt ($X_4 = 1$) into the equation from Table 10.6 yields the content of Melflux 2651f polycarboxylate superplasticizer, which will ensure the necessary concrete mixture workability of 13-15 cm.

Melflux 2651f = $0.41+0.095\cdot0.13\text{-}0.33\cdot(-1)+0.12\cdot(-1)+0.11\cdot1\text{-}0.06\cdot0.13^2$
$+0.2\cdot(-1)^2\text{-}0.03\cdot(-1)^2\text{-}0.02\cdot1^2\text{-}0.07\cdot0.13\cdot(-1)\text{-}$
$0.02\cdot0.13\cdot(-1)+0.04\cdot0.13\cdot1\text{-}0.05\cdot(-1)\cdot(-1)+0.04\cdot(-1)\cdot1$
$+0.04\cdot(-1)\cdot1 = 0.93\%$ by cement weight.

7. The aggregates consumption is calculated according to Eqs. 12.3-12.6:

$$V_a = 1000 - \left(\frac{507}{3.1} + \frac{152}{1} + \frac{80}{7.8} + \frac{4}{2.65}\right) = 672.7 \text{ l}$$

$$m_a = V_a\cdot\rho_a = 672.7 \times 2.7 = 1816 \text{ kg/m}^3$$

$$m_s = m_a \cdot 0.45 = 1816 \cdot 0.45 = 817 \text{ kg/m}^3$$
$$m_{c.s} = m_a \cdot 0.55 = 1816 \cdot 0.55 = 999 \text{ kg/m}^3$$

The calculated concrete has the following composition: cement – 507 kg/m^3, water – 152 l/m^3, crushed stone fractions 2-5 mm – 999 kg/m^3, sand – 817 kg/m^3. Content of Melflux 2651f superplasticizer is 0.93% cement weight, steel fibre content is 80 kg/m^3, basalt fibre is 4 kg/m^3.

The calculated of fibre-reinforced concrete composition should be verified experimentally.

Fibre-reinforced Concrete Composition Design According to the Minimum Cost Criterion

When designing concrete compositions, the main criteria for their optimisation are usually minimum cement consumption or minimum possible cost of concrete. For ordinary concrete, these criteria usually coincide. In both cases, it is mandatory to ensure a set of standardised properties of the concrete mixture and hardened concrete.

Modern concrete is a multi-component system and the cost of its individual components can approach or exceed the cement cost. An example of such concretes is fibre-reinforced concrete. It differs in the presence of at least three components (cement, fibres and plasticizing admixtures), the contents of which can vary in a wide range and have the main effect on the fibre-reinforced concrete's total cost. For fibre-reinforced concrete with two types of fibre, there are four such components. Thus, the task of fibre-reinforced concrete composition design aimed at minimising its cost is significantly complicated when using the traditional approach.

For example, Fig. 12.2 demonstrates the case when the required flexural strength of 17 MPa can be achieved with the same basalt fibre content of 4 kg/m^3, but with different W/C (0.3 and 0.4), different steel fibre contents (80, 100 and 120 kg), which respectively lead to different cement consumption (507, 516 and 472 kg). If we also take into account the need to determine the superplasticizer content, which affects both the concrete properties and its cost, as well as the possibility of providing another quality indicator (for example, compressive strength), then it becomes clear that it is practically impossible to solve the fibre-reinforced concrete composition optimisation problem by the traditional and including the described above nomographic method.

To solve fibre-reinforced concrete composition design problems with multivariate solutions, it is more effective to use mathematical programming methods [49, 97].

The condition of the problem for finding the optimal fibre-reinforced concrete composition with given quality indicators can be formulated as follows. Find the factors of fibre-reinforced concrete composition x_1-x_n, which allow to minimise its cost (CS_0):

$$CS_0 = CS_c \cdot C + CS_{ad} \cdot Ad + CS_F \cdot F \rightarrow min \qquad (12.7)$$

and providing the necessary quality indicators

$$I_1 \geq f(x_1, x_2, ..., x_n) \qquad (12.8)$$

$$I_2 \geq f(x_1, x_2, ..., x_n)$$

$$\cdots\cdots\cdots\cdots\cdots\cdots$$

$$I_m \geq f(x_1, x_2, ..., x_n)$$

$$\text{at } x_1 \ ... \ x_n \ [a...b] \qquad (12.9)$$

where CS_c, CS_{ad}, CS_F are respectively, the cost of cement, modifying admixture (superplasticizer, active mineral admixture, etc.) and fibre, per kg; C, Ad, F are accordingly, cement consumption, contents of modifying admixtures and fibre kg/m³ of fibre-reinforced concrete; I_1-I_m are specified fibre-reinforced concrete quality indicators; x_1-x_n are composition factors; a and b are constraints of the possible factors values.

In order to calculate the optimal composite fibre-reinforced concrete composition for the considered above example, it is necessary to solve the following mathematical programming problem:

Find such fibre-reinforced concrete mixture composition that would ensure the necessary compressive and flexural strength at 28 days with a minimum total cost within the permissible values of the factors.

The most rational way to solve such a problem is to use the Microsoft Excel software, in particular its 'Solver' unit, which is aimed to find solutions to equations and optimisation problems.

The design includes the following steps. Substitute the strength values that should be provided in the compressive and flexural strength models (Table 10.6) and the fibre-reinforced concrete components costs in Eq. (12.7). In Eq. (12.9), set the limits of the factors values (in coded form from –1 to 1). Next, the programme goes through various combinations of factors, providing the specified strength values according to Eqs. (12.8, 12.9), while minimising the function (12.7). To find the fibre-reinforced concrete cost during iterations, the necessary superplasticizer content is determined in parallel according to the corresponding equation (Table 10.6) with the intermediate values of factors x_1-x_4.

The result of such iterations is finding of the optimal values of composition factors: W/C, cement consumption, contents of steel and basalt fibre, as well as superplasticizer. Water demand and aggregate content can be calculated by Eqs. (12.3-12.6).

Example 2: *Design composition of fine-grained fibre-reinforced concrete with compressive strength at 28 days of 70 Mpa has flexural strength of 17 MPa, mixture slump of 130-150 mm by using experimental and statistical models (Table 10.6). The assumed cost of the main fibre-reinforced concrete components per kg* is: $CS_c = 3$ \$; $CS_{SF} = 50$ \$; $CS_{BF} = 90$ \$; $CS_{SP} = 260$ \$. *Materials: Portland cement M500, average quality fine aggregate with a fineness modulus* $M_f = 3.5$ *and real density* $\rho_a = 2.7$ *kg/l. The use of Melflux 2651f superplasticizer is assumed.*

1. By substituting the compressive and flexural strength values into the corresponding expressions (Table 30), obtain the constraint functions (2.22) of the problem:

$$78.4+5X_1-14.2X_2-0.1X_3+X_4-1.1X_1^2-0.6X_2^2-0.7X_3^2-$$
$$1.3X_4^2-0.1X_1X_2+0.2X_1X_3-0.35X_2X_3 \geq 70;$$
$$17.85+0.66X_1-2.03X_2+2.32X_3+X_4+0.88X_1^2+0.33X_2^2-1.62X_3^2$$
$$-0.57X_4^2-0.75X_1X_2-0.18X_1X_3-0.1X_1X_4-0.19X_3X_4 \geq 17$$

2. Substitute the fibre-reinforced concrete components cost in Eq. (12.7) and set the factor values limit from −1 to 1 (in coded form).
3. Using the 'Solver' unit, find the values of the factors that satisfy the problem constraints and minimise the total cost of fibre-reinforced concrete:

$$x_1 = 1;\ x_2 = -0.238;\ x_3 = -0.459;\ x_4 = -1$$

For such values of the factors according to the corresponding models (Table 10.6) $f_{c,tf}^{28} = 17$ MPa, which correspond to the required value of flexural strength of $f_{cm}^{28} = 83.0$ MPa, which is higher than the required compressive strength value.

4. The values of factors in their natural form are obtained from Eq. (12.1):

$$C = 50x_1 + 500 = 50 \cdot (1) + 500 = 550 \text{ kg/m}^3$$

$$W/C = 0.05x_2 + 0.35 = 0.05 \cdot (-0.238) + 0.35 = 0.338$$

$$SF = 20x_3 + 100 = 20(-0.459) + 100 = 90.8 \text{ kg/m}^3$$

$$BF = 2x_4 + 2 = 2(-1) + 2 = 0$$

5. The superplasticizer content according to the corresponding model (Table 10.6) is:

• in % of the cement weight:

$$SP' = 0.41+0.095X_1-0.33X_2+0.12X_3+0.11X_4-0.06X_1^2+0.2X_2^2-$$
$$0.03X_3^2-0.02X_4^2-0.07X_1X_2-0.02X_1X_3+0.04X_1X_4-$$
$$0.05X_2X_3+0.04X_2X_4+0.04X_3X_4 = 0.456\%$$

• by weight:

$$SP = SP' \cdot C/100 = 0.456 \cdot 550/100 = 2.51 \text{ kg/m}^3$$

6. The minimum possible cost of 1 m³ of fibre-reinforced concrete without taking into account the cost of aggregate and water (found during iterations using the 'Solver' unit according to Eq. (12.7)):

$$CS = 3 \cdot 550 + 260 \cdot 2.51 + 50 \cdot 90.8 = 6842.7 \text{ \$}$$

7. Water demand by Eq. (2.16) is:

$$W = 550 \cdot 0.338 = 185.9 \, 1$$

8. Aggregate content by Eq. (12.3):

$$A = \left(1000 - \left(\frac{550}{3.1} + \frac{90.8}{7.85} + \frac{177.5}{1}\right)\right) \cdot 2.7 = 1687 \, \text{kg}$$

The final composition of the fibre-reinforced concrete mixture, kg/m^3:

$$C = 550; \, W = 147; \, A = 1687; \, SF = 90.8; \, BF = 0; \, SP = 2.51.$$

The optimal composition design of composite fibre-reinforced concrete by the nomographic method would look as follows (Fig. 12.1, red line). It is obvious that it is impossible to solve such a problem by comparing and sorting many options manually.

At the task formulation stage for fibre-reinforced concrete composition design, it is necessary to correctly set the desired values of compressive and flexural strengths. It is obvious that these values should be within the minimum and maximum possible value of the initial parameter, since within these limits, the polynomial model adequately describes the investigated features. Such values can be found quite easily using the already mentioned 'Solution Search' unit. Thus, for the considered Example 2, the limit values of strengths within the factor variation region will be as follows:

$$f_{cm}{}^{28}(\text{min}) = 55.8 \text{ MPa}; f_{cm}{}^{28}(\text{max}) = 96.3 \text{ MPa};$$

$$f_{c,tf}{}^{28}(\text{min}) = 10.5 \text{ MPa}; f_{c,tf}{}^{28}(\text{max}) = 23.6 \text{ MPa}.$$

It is also possible to go beyond the output parameter limits. In this case, along with the optimisation problem, an extrapolation problem, allowing taking the factors' values outside the variation range, is also solved (for example, x_1-x_3 = 1.1; 1.2; 1.3). However, it is necessary to keep in mind that extrapolation may be associated with certain errors, which become more noticeable the further one goes beyond the variation range. Extrapolation is possible if, based on the research results, there is no doubt that the nature of the function remains unchanged outside the factor variation range.

Let's consider another example, which includes fixing one of the factors at a certain level.

Example 3: *The problem conditions fully correspond to the initial data of Example 2, but additionally it is necessary to limit the steel fibre content to the minimum possible value, i.e. SF = 80 kg/m^3.*

1. The constraint functions of the problem are similar to item 1 (Example 2).
2. Set the factor values limits: $x_1 = x_2 = -1$-1; $x_3 = -1$ (in coded form). Substitute Substitute the fibre-reinforced concrete components cost into Eq. 12.7.

3. Using the 'Solver' unit, find the values of factors that satisfy the problem constraints and minimise the total cost of fibre-reinforced concrete:

$$x_1 = 1; x_2 = -0.652; x_3 = -1; x_4 = -0.37.$$

For such factors values, according to the corresponding expressions (Table 10.6) $f_{c.tf}^{28} = 17$ MPa, which corresponds to the required flexural strength value, $f_{cm}^{28} = 89.7$ MPa, provide the required compressive strength value.

4. The factors values in their natural form are obtained by Eq. 12.1:

$$C = 50{\cdot}x_1 + 500 = 50{\cdot}1 + 500 = 550 \text{ kg}$$
$$W/C = 0.05x_2 + 0.35 = 0.05(-0.652) + 0.35 = 0.317$$
$$SF = 20x_3 + 100 = 20(-1) + 100 = 80 \text{ kg}$$
$$BF = 2x_4 + 2 = 2(-0.37) + 2 = 1.26 \text{ kg}$$

5. The superplasticizer content according to the corresponding expression (Table 10.6):
   - in % of cement weight: $SP' = 0.749\%$;
   - by weight: $SP = SP'{\cdot}C/100 = 0.749{\cdot}550/100 = 4.12$ kg.

6. The minimum possible cost of 1 m³ of fibre-reinforced concrete, excluding the cost of aggregate and water is:

$$W_{fc} = 3{\times}550 + 260{\times}4.12 + 50 \times 80 = 7048.4 \text{ \$}$$

7. The water demand following Eq. 2.16 is:

$$W = 550{\times}0.317 = 174.6 \text{ l.}$$

8. Following Eq. 12.3, the aggregate content is:

$$A = \left(1000 - \left(\frac{550}{3.1} + \frac{80}{7.85} + \frac{1.26}{2.7} + \frac{174.6}{1}\right)\right){\cdot}2.7 = 1721 \text{kg}$$

The final fibre-reinforced concrete composition, kg/m³ is:

$$C = 550; W = 175; A = 1721; SF = 80; BF = 1.26; SP = 4.12.$$

Analysing the resulting composition and comparing it with that obtained in Example 2, it should be noted that when limiting the steel fibre content, it becomes more difficult to achieve the specified flexural strength and, therefore, it is necessary to overuse cement and superplasticizer as it leads to formation of a significant reserve in terms of compressive strength – 89.7 MPa, instead of the required 70 MPa.

The proposed fibre-reinforced concrete composition design method allows considering the specific features of the investigated materials and to optimise the composition quite easily according to a given criterion, for example, the minimum cost. Additionally, one of the advantages of the method is the possibility of setting an arbitrary number of constraints, which allow providing simultaneously a significant number of quality indicators, which can be neither higher nor less than the given value.

Conclusion

The monograph provides an overview of theoretical ideas about the influence of the concrete mixtures composition factors in the production of high-strength rapid-hardening concrete and technological ways to ensure them. Calculation and experimental bases of cement hydration degree influence on the cement stone strength and its change over time at extremely low values of water-cement ratio and various specific surface area of cement are given. The influence of various types of plasticising admixtures and their water-reducing ability on increasing the early strength of cement stone and concrete is shown. As an alternative to increasing the cement specific surface area, the possibility of using calcium nitrate as hardening accelerating admixture is considered.

It has been found that the strength growth kinetics of cement stone on finely ground alite cement with addition of hardening accelerator differs significantly from the traditional one and is characterised in the latter case at $W/C = 0.2$-0.3 by an increase in compressive strength after 12 h of hardening up to 50%, and after one day, up to 70% of the strength value at 28 days. Adding complex admixtures of Melflux 2641F polycarboxylate superplasticizer and hardening accelerator into the concrete mixture made it possible, at $W/C = 0.25$ and concrete mixture slump of 160-220 mm, to achieve concrete strength of 55-58 MPa at 12 h for a 28-day strength of 100-115 MPa. Long-term tests of concrete (up to 180 days) did not show a noticeable decrease in its compressive strength.

To take into account the joint influence of the main factors of high-strength rapid-hardening concrete composition on its strength, a set of experimental-statistical models was obtained using mathematical planning.

The monograph presents algorithms for obtaining and analysing experimental-statistical models of strength and other properties of high-strength rapid-hardening concrete and ways for design on their basis optimal concrete mixtures compositions.

A number of specific examples were solved. Calculated dependencies, reflecting the joint effect on concrete compressive strength at different ages, along with *W/C*, consumption and strength of cement, both under normal hardening conditions and under various modes of heat and moisture treatment, were obtained. Along with high-strength concrete compressive strength, polynomial models of splitting tensile strength were obtained for the case when composite ash and slag cements were used. This enabled to expand the range of tasks for design concrete compositions, including, along with the required compressive strength value, the normalised value of splitting tensile strength.

A problem of obtaining and analysing experimental-statistical models that reflect the aggregate grain composition influence on the water demand and high-strength concrete strength is considered. The resulting models allow predicting the concrete strength and water demand for a given aggregate grain composition and also to determine the necessary fractions ratio to ensure the required concrete properties.

To design concrete mixtures compositions with required values of strength parameters and workability, experimentally statistical models of concrete mixtures water demand were obtained, taking into account the given cone slump. These models enable to study how the mixture water demand depends on the necessary cone slump, superplasticizer content and cement specific surface area and consumption.

In a separate chapter, the monograph deals with problems of concrete compositions multi-parameter design, in which, along with the required strength parameters, other concrete properties, determining its deformability, frost and water resistance, are considered. The necessary design equations and models of the specified properties and algorithms for their joint use in corresponding tasks are proposed. Examples of solving problems for high-strength concrete compositions optimal multi-parametric design are presented.

The second chapter of the monograph deals with results of experimental studies on dispersed-reinforced high-strength concrete (fibre-reinforced concrete). Using the obtained experimental-statistical models, the influence on compressive and flexural strength of high-strength ordinary concrete with crushed granite with a particle size of 5-20 mm and fine-grained concrete with a limiting aggregate size of 5 mm is considered. The models analysis shows the undoubted advantage of using corrugated fibre, which can be explained by its increased adhesion surface with the fibre-reinforced concrete matrix component. The advantages of corrugated fibre are most evident for flexural strength.

As experiments have shown, along with the composition, adding fibres into the concrete mixture has a significant impact on fibre-reinforced concrete strength characteristics. Fibre clumping and a corresponding decrease in strength parameters were noted with a mixture workability of up to 150 mm. For higher workability provided by adding superplasticizers into the mixture was not noted. A positive effect, as shown by experiments, is achieved by orienting steel fibres in concrete mixtures with high workability under the influence of a magnetic field during strength indicators and were determined at one, seven and 28 days, taking into account the water-cement ratio and cement consumption for concrete reinforced with hooked and corrugated fibres vibration.

Along with the high-strength fibre-reinforced concrete strength characteristics, the monograph discusses the results of experimental studies of their properties, which significantly affect the concrete durability during its lifetime. The influence of the water-cement ratio and the water content of concrete mixtures with different fibre content on the parameters of the concrete pore structure is considered. Adding superplasticizers has a positive effect on these parameters.

For various fibre-reinforced concrete compositions, shrinkage deformations, frost resistance, residual flexural strength were studied and the optimal conditions were determined to achieve their required values.

Along with studying of the influence of technological parameters on the high-strength steel fibre-reinforced concrete properties, the monograph presents the results for high-strength fibre-reinforced concrete with composite dispersed reinforcement. The influence of steel, basalt and polypropylene fibres consumption and their ratio on the fibre-reinforced concrete strength in tension and compression has been determined. The best results were obtained with a combination of steel and basalt fibres when the content of the latter was in the range of 2-4 kg/m^3 and steel fibre content varied within 80-120 kg/m^3.

An important advantage of fibre-reinforced concrete is the increased dynamic characteristics. The monograph presents the results of experimental studies on the effect of fibre-reinforced concrete composition on the impact strength. It is shown that the maximum work value required for destruction of specimens under impact, referred to their volume, is achieved with composite dispersed reinforcement using steel and basalt fibres.

The experimental-statistical models obtained in the research enabled to propose the methodology outlined in the monograph for design of optimal high-strength fibre-reinforced concrete compositions, providing a set of their normalised properties at the lowest possible cost.

References

1. Feret, R., *Technology of Building Binders*, 1902, St. Peterburg, p. 132 (in Russian).
2. Sheikin, A.E., *Structure, Strength and Crack Resistance of Cement Stone*, 1974, Moscow, Stroyizdat, p. 192 (in Russian).
3. Itskovich, S.M., Chumakov, L.D. and Bazhenov, Y.M., *Technology of Concrete Aggregates*, 1991, Moscow, Vysshayashkola, p. 272 (in Russian).
4. Itskovich, S.M., *Large-pore Concrete (Technology and Properties)*, 1977, Moscow, Stroyizdat, p. 119 (in Russian).
5. Dvorkin, L. and Dvorkin, O., *Basics of Concrete Science*, 2006, St. Petersburg, Stroybeton, p. 686 (in Russian).
6. Powers, T., The physical structure of cement and concrete, *Cement and Lime Manufacture*, 1956, 29(2): 270.
7. Dvorkin, L.I. and Dvorkin, O.L., *Computational Prediction of Properties and Design of Concrete Compositions*, Educational and Practical Guide, 2016, Moscow, Infra-Inzheneriya, p. 384 (in Russian).
8. Skramtaev, B.G., Shubenkin, P.F. and Bazhenov, U.M., *Methods for Determining the Composition of Concrete of Various Types*, 1966, Moscow, Stroyizdat (in Russian).
9. Bazhenov, Y.M., *Concrete Technology*, 1987, Moscow, Vysshayashkola, p. 449 (in Russian).
10. Sheykin, A.E., Chechovskiy, Yu.V. and Brusser, M.I., *Structure and Properties of Cement Concretes*, 1979, Moscow, Stroyizdat, p. 344 (in Russian).
11. Sizov, V.P., *Design of Normal-weight Concrete Compositions*, 1980, Moscow, Stroyizdat, p. 144 (in Russian).
12. Kaiser, L.A. and Chekhova, R.S., *Cements and Their Rational Use in the Production of Prefabricated Reinforced Concrete Products*, 1972, Moscow, Stroyizdat, p. 80 (in Russian).
13. Ahverdov, I.N., *Basis of Concrete Physics*, 1981, Moscow, Stroyizdat, p. 464 (in Russian).
14. Grushko, I.M., Ilyin, A.G. and Chikhladze, Ye.D., *Tensile Strength of Concretes*, 1973, Kharkov, Publishing House of KhGU, p. 155 (in Russian).
15. Mikhailov, N., *Basic Principles of the New Technology of Concrete and Reinforced Concrete*, 1961, Moscow, Gosstroyizdat, p. 52 (in Russian).

16. Zaporozhets, I.D., Okorokov, S.D. and Pariyskiy, A.A., *Heat Release of Concrete*, 1966, Leningrad-Moscow, Stroyizdat, p. 314 (in Russian).
17. Shmigalskiy, V.N., *Optimisation of Cement Concrete Compositions*, 1981, Chisinau, Shtinca, p. 123 (in Russian).
18. Kostuch, J.A., Walters, G.V. and Jones, T.K., High Performance Concrete Incorporating Metakaolin – A Review, Concrete 2000 Conference, September 1993, University of Dundee.
19. BS EN 206:2013+A1:2016, Concrete, Specification, performance, production and conformity.
20. Nawy, E.G., *Fundamentals of High-strength High-performance Concrete*, 1996, Harlow, Longman Group Limited, p. 360.
21. Dilsa, J., Boelb, V. and De Schuttera, G., Influence of cement type and mixing pressure on air content, rheology and mechanical properties of UHPC, *Construction and Building Materials*, 2013, 41: 455-463.
22. Kerkhoff, B., Benefits of Air Entrainment in HPC, *HPC Bridge Views*, 2002, 23(3).
23. Aïtcin, P.C., *High Performance Concrete* (1st ed.), 1998, CRC Press, https://Doi.org/10.4324/9780203475034.
24. Dvorkin, L. and Dvorkin, O., *Basics of Concrete Science: Optimum Design of Concrete Mixtures*, 2006, Kindle Edition, p. 237.
25. Gridchin, A.M., Features of the production of binders of low water demand and concrete based on it using technogenic polymineral sand, *Building Materials, Equipment, Ttechnologies of the XXI Century*, 2002, 1(36) (in Russian).
26. Batrakov, V., Superplasticisers – Research and experience of application, *Application of Chemical Additives in Concrete Technology*, 1980, Moscow, Knowledge, 29-36 (in Russian).
27. Beushausen, H. and Dittmer, T., The influence of aggregate type on the strength and elastic modulus of high strength concrete, *Construction and Building Materials*, 2015, 74: 132-139.
28. Olginsky, A.G. and Melnik, Y.M., Concrete on activated aggregate, *Management of Structure Formation, Structure and Properties of Road Concrete*, 1983, Kharkov, 137-138 (in Russian).
29. Solomatov, V.I., Borbyshev, A.P. and Proshin, N.P., Clusters in the structure and technology of composite building materials, *Izv. Universities, Construction and Architecture*, 1983, 4: 56-61 (in Russian).
30. ASTM C494/C494M – 19, *Standard Specification for Chemical Admixtures for Concrete*.
31. Ramachandran, V.S., *Concrete Admixtures Handbook: Properties, Science and Technology*, 2nd ed., 1995, New Jersey, Noyes Publications, p. 1183.
32. Doroshenko, Y.M., Vishnevsky, V.B. and Chistyakov, V.V., Additives that increase the strength and water resistance of cement concrete, *Improving the Durability of Water Management Structures*, Abstracts of the reports of the All-Union Conference, Rostov-on-Don, 1981, 133-135 (in Russian).
33. Powers, T.C. and Brownyard, T.L., Studies of the physical properties of hardened Portland cement paste, *J. Am. Concrete Inst. Proc.*, 1947, 43: 101-132.
34. Dvorkin, L., *Optimal Concrete Composition Design*, 1981, Lviv, Vyshcha Shkola, p. 159 (in Ukrainian).
35. Locher, F.W., *Cement Principles of Production and Use*, 2006, Dusseldorf: Verlaq Bau+Technic YmbH, p. 535.

36. Batrakov, V.G., *Modified Concrete: Theory and Practice*, 2nd edition, 1998, Moscow, Tekhnoproekt, p. 768 (in Russian).
37. Ratinov, V.B. and Rozenberg, T.M., *Admixtures in Concrete*, 1989, Moscow, Stroyizdat, p. 188 (in Russian).
38. Kravchenko, I.V., Vlasova, M.T. and Yudovich, B.E., *High-strength and Especially Fast-hardening Portland Cements*, 1971, Moscow, Stroyizdat, p. 233 (in Russian).
39. Volzhensky, A.V., *Mineral Binders*, 1973, Moscow, Stroyizdat, p. 479 (in Russian).
40. Bazhenov, Y.M., *Modified High-quality Concrete*, 2006, Moscow, Association of Civil Engineering Education, p. 368 (in Russian).
41. Montgomery, D.C., *Design and Analysis of Experiments*, 5th ed., 2000, New Jersey: Wiley, p. 688.
42. Dvorkin, L., Dvorkin, O. and Ribakov, Y., *Mathematical Experiments Planning in Concrete Technology*, 2012, New York, Nova Science Publishers, Inc, p. 173.
43. Box, G.E.P., Hunter, J.S. and Hunter, W.G., *Statistics for Experimenters: Design, Discovery and Innovation*, 2nd ed., 2005, New Jersey, Wiley, p. 672.
44. Dvorkin, L.I., Solomatov, V.I., Vyrovoj, V.N. and Chudnovskij, S.M., *Cement-based Concrete with Mineral Fillers*, 1991, Kyiv, Budivelnyk, p. 136 (in Russian).
45. EN 206-1:2000, *Concrete*, Part 1: *Specification, Performance, Production and Conformity*, CEN.
46. EN 12390-1:2021, *Testing Hardened Concrete*, Part 1: *Shape, Dimensions and Other Requirements for Specimens and Moulds*, CEN.
47. Hewlett, P. (Ed.), *Lea's Chemistry of Cement and Concrete*, 4th edition, 2003, Imprint: Butterworth Heinemann, p. 1092.
48. Neville, A.M., *Properties of Concrete*, 4th edition, 1996, New York, Wiley & Sons, p. 844.
49. Dvorkin, L., Bordiuzhenko, O., Zhitkovsky, V. and Marchuk, V., Mathematical modelling of steel fibre-reinforced concrete properties and selecting its effective composition, *IOP Conference Series: Materials Science and Engineering*, 2019, 708(1): 012085.
50. Kuneva, V., Milev, M. and Gocheva, M., Modelling the transportation assessment with MS excel solver, *AIP Conference Proceedings*, 2021, 2333, 150005, https://Doi.org/10.1063/5.0042520.
51. Shah, A.A. and Ribakov, Y., Recent trends in steel fibred high-strength concrete, *Mater. Des.*, 2011. 32: 4122-4151, Doi:10.1016/j.matdes.2011.03.030.
52. Thomas, J. and Ramaswamy, A., Mechanical Properties of Steel Fibre-reinforced Concrete, *ASCE J. Mater. Civ. Eng.*, 2007. 19: 385-392. Doi:10.1061/(ASCE)0899-1561(2007)19: 5(385).
53. Klyuyev, S.V., High-strength fibre concrete for industrial and civil construction, *Magazine of Civil Engineering*, 2012, 8: 61-66.
54. Beaudoin, J.J., *Handbook of Fibre-reinforced Concrete: Principles, Properties, Developments and Applications*, 1990, Building Materials Science, Noyes, William Andrew.
55. Maidl, B., *Steel Fibre-reinforced Concrete*, 1995, Wiley, Ernst & Sohn.
56. EN 14889-1:2006, *Fibres for Concrete*, Part 1: *Steel Fibres – Definitions, Specifications and Conformity*.
57. Doyon-Barbant, J. and Charron, J.P., Impact of fibre orientation on tensile, bending and shear behaviours of a steel fibre-reinforced concrete, *Mater Struct.*, 2018, 51: 157. https://Doi.org/10.1617/s11527-018-1282-0

58. Achilleos, C., Hadjimitsis, D., Neocleous, K., Pilakoutas, K., Neophytou, P. and Kallis, S., Proportioning of steel fibre-reinforced concrete mixes for pavement construction and their impact on environment and cost, *Sustainability*, 2011, 3(7): 965-983, https://Doi.org/10.3390/su3070965

59. Wang, J., Niu, D. and Zhang, Y., Mechanical properties, permeability and durability of accelerated shotcrete, *Construction and Building Materials*, 2015, 95: 312-328.

60. Klyuev, S.V., Khezhev, T.A., Pukharenko, Yu.V. and Klyuev, A.V., Experimental study of fibre-reinforced concrete structures, *Materials Science Forum*, 2019, 945, 115-119, https://doi.org/10.4028/www.scientific.net/msf.945.115

61. Torrents, J.M., Blanco, A., Pujadas, P., Aguado, A., Juan-Garcia, P. and Sanchez-Moragues, M.A., Inductive method for assessing the amount and orientation of steel fibres in concrete, *Mater Struct.*, 2012, 45: 1577-1592.

62. Boulekbache, B., Hamrat, M., Chemrouk, M. and Amziane, S., Influence of yield stress and compressive strength on direct shear behaviour of steel fibre-reinforced concrete, *Construction and Building Materials*, 2012, 27(1): 6-14.

63. Kaufmann, J., Frech, K., Schuetz, P. and Münch, B., Rebound and orientation of fibres in wet sprayed concrete applications, *Construction and Building Materials*, 2013, 49: 15-22.

64. Mu, R., Li, H., Qing, L., Lin, J. and Zhao, Q., Aligning steel fibres in cement mortar using electro-magnetic field, *Construction and Building Materials*, 2017, 131: 309-316.

65. Javahershenas, F., Gilani, M.S. and Hajforoush, M., Effect of magnetic field exposure time on mechanical and microstructure properties of steel fibre-reinforced concrete, *Journal of Building Engineering*, 2021, 35: 101975.

66. Guidance for the design of steel-fibre-reinforced concrete, 2007, *Concrete Society Technical Report*, pp. 92.

67. Gul, M., Bashir, A. and Naqash, J.A., Study of modulus of elasticity of steel fibre-reinforced concrete, *Int. J. Eng. Adv. Technol.*, 2014, 3(4): 304-309.

68. Carrillo, J., Ramirez, J. and Lizarazo-Marriaga, J., Modulus of elasticity and Poisson's ratio of fibre-reinforced concrete in Colombia from ultrasonic pulse velocities, *Journal of Building Engineering*, 2019, 23: 18-26.

69. Faisal, F.W. and Ashour, S.A., Mechanical properties of high-strength fibre-reinforced concrete, *ACI Material Journal*, 1992, 89(5): 449-455.

70. Swamy, R.N. and Mangat, P.S., Influence of fibre geometry on the properties of steel fibre-reinforced concrete, *Cement and Concrete Research*, 1974, 4(3): 451-465.

71. Klyuev, S.V., Khezhev, T., Pukharenko, Y.V. and Klyuev, A.V. (2018). The fibre-reinforced concrete constructions experimental research. *In*: *Materials Science Forum*, 931, pp. 598-602, Trans Tech Publications Ltd.

72. Jo, B.W., Shon, Y.H. and Kim, Y.J., The evaluation of elastic modulus for steel fibre-reinforced concrete, *Russian Journal of Non-destructive Testing*, 2001, 37(2): 152-161.

73. Kizilkanat, A.B., Kabay, N., Akyüncü, V., Chowdhury, S. and Akça, A.H., Mechanical properties and fracture behavior of basalt and glass fibre-reinforced concrete: An experimental study, *Construction and Building Materials*, 2015, 100: 218-224.

74. El-Newihy, A., Azarsa, P., Gupta, R. and Biparva, A., Effect of polypropylene fibres on self-healing and dynamic modulus of elasticity recovery of fibre-reinforced concrete, *Fibres*, 2018, 6(1): 9.

75. Gilles, D., Ziad, H., Alain, S., Sandriue, C. and Luc, W., Precast thin UHPFRC curved shells in a waste water treatment plant, *RILEM-fib-AFGC Int. Symposium on Ultra-High Performance Fibre-Reinforced Concrete*, 2013, France, Marseille, 49-58.
76. Kolsky, H., An investigation of the mechanical properties of material at very high rates of loading, *Proc. Phys. Soc.* (London), 1949, 62B: 676-700.
77. Naaman, A.E., Engineered steel fibres with optimal properties for reinforcement composites, *Journal of Advanced Concrete Technology*, 2003, 1: 241-252.
78. *Recommendations for Design and Construction of Ultra-high Strength Fibre-reinforced Concrete Structures*, Concrete Committee of Japan Society of Civil Engineers, 2006.
79. Rabinovich, F.N., Composites based on dispersed reinforced concrete, *Issues of Theory and Design, Technology, Design*, 2004, Moscow (in Russian).
80. STO NOSTROY 2.27.125-2013. Structures of transport tunnels made of fibre-reinforced concrete, *Rules for the Design and Production of Works*, 2012, Moscow (in Russian).
81. Sheikin, A.E., *Structure, Strength and Crack Resistance of Cement Stone*, 1974, Moscow, Stroyizdat (in Russian).
82. DSTU B V.2.7-170:2008, Methods of determination of middle density, moisture content, water absorptions, porosity and water tightness, *Building Materials: Concretes*, 2009, Kyiv (Ukrainian standard).
83. Nizina, T.A., Balykov, A.S., Volodin, V.V. and Korovkin, D.I., Fibre fine-grained concretes with polyfunctional modifying additives, *Magazine of Civil Engineering*, 2017, 4(72): 73-83.
84. Whiteside, T. and Sweet, H., *Proceedings of Highway Research Board*, 1950, 30, p. 204.
85. DSTU B V.2.7-49-96, Accelerated methods for determining frost resistance during multiple freezing and thawing, *Building Materials: Concretes*, 1996, Kyiv (Ukrainian standard).
86. Grote, D.L., Park, S.W. and Zhou, M., Dynamic behaviour of concrete at high strain-rates and pressures: I. Experimental characterisation, *Int. J. Impact Engng.*, 2001, 25: 869-886.
87. Soufeiani, L., Raman, S.N., Jumaat, M.Z.B., Alengaram, U.J., Ghadyani, G. and Mendis, P., Influences of the volume fraction and shape of steel fibres on fibre-reinforced concrete subjected to dynamic loading – A review, *Engineering Structures*, 2016, 124: 405-417.
88. Asprone, D., Cadoni, E., Prota, A. and Manfredi, G., Dynamic behaviour of a Mediterranean natural stone under tensile loading, *International Journal of Rock Mechanics and Mining Sciences*, 2009, 46(3): 514-520.
89. EN 14488-5:2006, *Testing Sprayed Concrete*, Determination of energy absorption capacity of fibre-reinforced slab specimens.
90. EN 14651:2005+A1:2007, Test method for metallic fibre concrete, *Measuring the Flexural Tensile Strength* (limit of proportionality (LOP), residual).
91. EN 14845-2:2006, *Test Methods for Fibres in Concrete: Effect on Concrete*. https://standards.iteh.ai/catalog/standards/cen/532667bb-b16a-4ae6-9f1d-d3149994c0e0/en-14845-2-2006
92. Dvorkin, L., Bordiuzhenko, O., Tekle, B.H. and Ribakov, Y., A method for the design of concrete with combined steel and basalt fibre, *Applied Sciences*, 2021, 11(19): 8850, https://Doi.org/10.3390/app11198850

93. Graf, O., Strength and elasticity of high-strength concrete, *Deutscher Ausschuss für Stahlbeton*, 1954, 113, Vereag Ernst und Sohn, Berlin (in German).
94. Brandt, A.M., *Optimisation Methods for Material Design of Cement-based Composites*, 1998, E&FN Spon, p. 328.
95. Neville, A.M., *Properties of Concrete*, 2000, Krakow, vol. 4, p. 874.
96. Naaman, A.E., Engineered steel fibres with optimal properties for reinforcement composites, *Journal of Advanced Concrete Technology*, 2003, 1: 241-252.
97. Dvorkin, L., Dvorkin, O. and Ribakov, Y., *Multi-parametric Concrete Compositions' Design*, 2013, New York, Nova Science Publishers, p. 223.

Index

For Product Safety Concerns and Information please contact our EU
representative GPSR@taylorandfrancis.com
Taylor & Francis Verlag GmbH, Kaufingerstraße 24, 80331 München, Germany